IN THE SHADOW OF THE BOMB

IN THE SHADOW OF THE BOMB:

Oppenheimer, Bethe, and the Moral Responsibility of the Scientist

S. S. Schweber

PRINCETON UNIVERSITY PRESS

PRINCETON AND OXFORD

Third printing, and first paperback printing, 2007
Paperback ISBN-13: 978-0-691-12785-9
Paperback ISBN-10: 0-691-12785-9

The Library of Congress has cataloged the cloth edition of this book as follows

Schweber, S. S. (Silvan S.)
In the Shadow of the Bomb: Oppenheimer, Bethe,
and the moral responsibility of the scientist /
S. S. Schweber
p. cm. – (Princeton series in physics)
Includes bibliographical references and index.
ISBN 0-691-04989-0 (cloth : alk. paper)
1. Oppenheimer, J. Robert, 1904–1967. 2. Bethe, Hans Albrecht, 1906–
3. Atomic bomb–Moral and ethical aspects–United States.
4. Nuclear physics–United States–Biography. I. Title. II. Series.

QC774.O56 S32 2000
172'.422–dc21   99–052225

British Library Cataloging-in-Publication Data is available

This book has been composed in Utopia and Bluejack display

Printed on acid-free paper. ∞

pup.princeton.edu

Printed in the United States of America

10 9 8 7 6 5 4

For Paul Forman and Anne Harrington,

special friends,

Everett Mendelsohn,

a special colleague,

and Miriam

Contents

Preface

I used to be a theoretical physicist. I did my graduate studies at Princeton University from 1949 to 1952. During my stay in Princeton, theoretical physics was dominated by Eugene Wigner, the Jones Professor of Mathematical Physics at the university, and Robert Oppenheimer, who recently had become the director of the Institute for Advanced Study. On Fridays, the departmental colloquium would bring Oppenheimer, when he was in town, to the university, and the comments by Wigner and Oppenheimer would expose their contrasting views of the world. Every Wednesday afternoon all the theory students would trek out to the Institute to attend the theoretical physics seminar there and listen to Oppenheimer's often acerbic comments on the presentation.

Princeton was an enormously stimulating place. I still vividly remember Einstein giving a series of lectures on unified field theory; Bohr and Einstein presenting their differing views on quantum mechanics in inaudible and incomprehensible mutterings; Pauli talking on spin and statistics; von Neumann commenting on David Bohm's formulation of the quantum theory of measurement and his walking out of a lecture by Julian Schwinger. The atmosphere in Fine Hall, the home of the mathematics department, was equally heady. In addition, there was a rich cultural life. I recall Hermann Weyl delivering his lectures on symmetry; Eugene Wigner giving a talk at the Graduate School on the limits of science; Bertrand Russell giving a lecture to a packed auditorium, and Dylan Thomas arriving late and drunk to give passionate, moving readings of his poetry.

But the years I spent at Princeton were also tense times. The Cold War had intensified, and when I arrived there in September 1949 the USSR had just detonated its first plutonium bomb. That fall, an intense debate took place on whether to develop an H-bomb. I caught my first glimpse of Hans Bethe when he was leaving Wigner's office sometime in October 1949. He was in Princeton to discuss with Oppenheimer the feasibility of a fusion bomb. The following spring, David Bohm, a brilliant young theoretical physicist who had been a staff member at the Radiation Laboratory at Berkeley during the war and then became an assistant professor in Princeton's Department of Physics, was cited in contempt of Congress; he had taken the Fifth Amendment when refusing to answer certain questions posed to him by the House Un-American Activities Committee. In 1950 he was indicted. When the president of Princeton University, Harold Dodds, decreed thereafter that Bohm could not set foot on campus, a delegation of physics graduate students, including me, went to speak to him to appeal that decision. After a brief exchange, we were reprimanded, reminded that "Gentlemen, there is a war on!" and were invited to leave. Two of us then went to see Oppenheimer, who had been Bohm's teacher at Berkeley, and he graciously offered Bohm a desk at the Institute.

In the fall of 1952, I went to Cornell as a postdoctoral fellow to work with Hans Bethe. Freeman Dyson, who had been appointed the previous year to a full professorship to replace Richard Feynman, who had gone to Cal Tech, was also there. I was one of a large contingent of postdoctoral fellows and research associates who, together with the half-dozen or so graduate students in theory, formed a lively intellectual community. All of us had offices in the recently built Newman Laboratory for Nuclear Studies, and every day nearly all of us would go out to lunch with Bethe—when he was in town—at the nearby cafeteria run by Cornell's Department of Home Economics.

On arriving at Cornell I was struck by the fact that the doors to the offices of Robert (Bob) Wilson, the director of the laboratory, and Bethe were always open. The professors' doors at Princeton had always been closed. Bethe and Wilson shared a secretary, Velma Ray, whose station was at the entrance to their offices.[1] The intensity and closeness of the interaction between theorists and experimenters were also immediately apparent. Every weekend there would be a party at the apartment of one of the postdoctoral fellows or research associates to which everyone was invited. And it was not unusual for people to get drunk on these occasions—liquor was plentiful and no one had any apprehensions about letting their guard down.

A pervasive sense of community is the strongest impression I have of Cornell. Among the high-energy theorists the sense of community was heightened by the fact that during the two years I spent at Cornell, most of us were engaged in a collective research program on meson-nucleon scattering. At the time, it was not clear to me who or what was responsible for the sense of community. I took for granted the various forums that melded the community. In the Nuclear Lab, we had weekly Friday afternoon gatherings at which theorists and experimenters presented their latest findings in a form that everyone would understand, and there were weekly Monday evening meetings of the Journal Club, at which the latest papers and preprints would be discussed critically for the benefit of everyone. In addition, every week on Monday afternoon, a colloquium brought together the high-energy staff of the Newman Lab and the other members of the department, most of whom were were solid state physicists.

I came to Brandeis in the fall of 1955 and have been there ever since. Although I became deeply involved in the building of this university, which was founded in 1948, and in particular with the creation of its physics department, I managed to keep up with developments in physics and to remain active in research until the mid-sixties. However, events during the second half of the decade proved disruptive. Quantum field theory—my field of research—fell into disrepute. The Vietnam conflict and the ensuing student unrest got me more deeply absorbed in campus politics.

When field theory came back in favor in the late sixties and early seventies, broken symmetry and nonabelian gauge theories were thrust center stage. For some reason, I had difficulty adjusting to the new viewpoints, and by the mid-seventies I found myself unproductive in physics. At that same time, the pressures connected with justifying the large size of the physics faculty had led the department to offer courses that would attract non-science majors. I had always been interested in such activities, and I volunteered to teach a course on the introduction of probability into the sciences. With the characteristic hubris of the physicist, I had always assumed that physicists had been responsible for the introduction of statistical and probabilistic methods into the sciences—think of Maxwell and of Boltzmann and the second law of thermodynamics. Darwin and eighteenth-century political economy proved me wrong. Reading Howard Gruber's *Darwin on Man* opened up a whole new world for me, and I decided to spend a sabbatical leave during 1976–77 in the Department of the History of Science at Harvard. I have been a historian of science ever since.

My first activities as a historian concerned the young Charles Darwin and the genesis of natural selection, and the young Auguste Comte and the formulation of the nebular hypothesis. Many of my historical researches have had a biographical component—with creativity as a latent theme. I have written articles on Charles Darwin, John Herschel, Auguste Comte, and John Slater. *QED and the Men Who Made It*, my book on the history of quantum electrodynamics, contains fairly extensive biographical accounts of Sin-Itiro Tomonaga, Julian Schwinger, Richard Feynman, and Freeman Dyson as well as shorter sketches of the lives of—among others—Paul Dirac, Pascual Jordan, Willis Lamb, Wolfgang Pauli, Norman Kroll, and Quin Luttinger. My writing about young, creative scientists was an attempt to better understand the roots of their creativity. Thus, my articles on Darwin were concerned with the context and the intellectual catalysts that led to his formulation of natural selection and the principle of divergence of character; the one on Feynman with an analysis of the genesis of Feynman diagrams; and my study on the parallel lives of John Herschel and Darwin probed the differences in their creative efforts. The subtext of the latter inquiry[2] was to try to understand better the constraints imposed—and the freedoms granted—by the communities and the society to which they belonged, as compared to the constraints of nature and its latitude. My concern with off-scale individuals undoubtedly stemmed from the fact that I was socialized into a culture that attributes the past successes of science to the accomplishments of great individuals, usually men, who had the ability to read nature better than others. Most of my scientific activities had been carried out in a field whose research program at any one time was conceived, formulated, and usually dominated by a single individual: Julian Schwinger, Richard Feynman, Freeman Dyson, Murray Gell-Mann, Stanley Mandelstam, Geoffrey Chew, Steven Weinberg, among others. Although my historical studies have placed heavy emphasis on the individual scientist, I have always stressed the importance of the social and institutional setting in which the scientific activities were carried out. And, when considering the psychological factors involved in the creativity of the theorists that I have studied, I have been sensitive about the social psychology of the theoretical physics community.

Ten years ago, Hans Bethe asked me to write his intellectual biography. The project has grown into a full biography. This has been an exhilarating and deeply rewarding enterprise, principally because of my association with him.

I have diligently collected and studied the materials historians amass when writing biography. To date, Bethe has published over three hundred scientific and technical papers. Many of these are lengthy review articles of the same quality and comprehensiveness as his early *Handbuch* articles and his seminal articles on nuclear physics in the *Reviews of Modern Physics*, now usually referred to as the Bethe *Bible*, which he wrote during the 1930s.[3] He has also authored a very large number of reports and papers that were originally classified. Some of these became classics upon declassification, even though they have never been published in the open literature. To give an indication of the magnitude of the corpus of these works, I obtained from the Los Alamos Archives a five hundred page report on shock waves that Bethe wrote during the war which I had requested to be declassified. It contains much that was original and new at the time it was drafted. Bethe has also written extensively on the issues connected with weapons proliferation, test ban treaties, anti-ballistic missiles (ABM), disarmament, nuclear power, and the Strategic Defuse Initiative (SDI). There are by now over sixty cartons of Bethe papers in the Cornell archives. They contain materials up to the late 1980s: correspondence, drafts of published and unpublished papers, calculations, lecture notes, class notes, and extensive notes on the numerous seminars and conferences he attended. Also in the Cornell archives are the records of his role as adviser to various government agencies, national laboratories (Los Alamos, Oak Ridge, Brookhaven), and those connected with his extensive consulting activities with industrial laboratories, including GE, Detroit Edison, and Avco.

The extensive correspondence between Bethe and his friend Rudolf Peierls; between Bethe and his teacher and *Doktorvater* Arnold Sommerfeld; between Bethe and Victor Weisskopf, Isidor Rabi, J. Robert Oppenheimer, and Richard Feynman—all is on deposit in various archives. His former students have kept the letters they received from him, and the record of their interactions often fills many notebooks: there are over seventy graduate students who have completed dissertations under Bethe's guidance. In addition, there is an extensive extant correspondence between Bethe and his mother, and since 1968 he has kept a detailed diary. Many more papers are to be deposited in the Cornell archives, as Bethe has remained active in research and in the public sphere.

One is thus overwhelmed by documents, letters, and papers. It is clearly impossible for me to master all this wealth of information in the finite amount of time available. When three years ago I looked at what I had already written and what had yet to be done, it became apparent that the biography would become three fat volumes that not many people would read. I felt strongly that this would be unfair to Bethe, for I believe his life and his accomplishments ought to become better known. I then decided to try to write a book of some five to six hundred pages in length that would be accessible to a wide audience.

The challenge in writing the biography thus became selection. To give a more balanced presentation I decided to write a good deal of the biography as studies in parallel lives. Thus, the chapter on energy generation in stars—the work done in the late 1930s for which Bethe won the Nobel Prize in 1967—is a study of George Gamow's and Hans Bethe's contributions to the formulation and the solution of the problem and contrasts their "styles." Another chapter compares the life of Rudolf Peierls with that of Bethe to illustrate contingency due to the heterogeneity of contexts. It asks the question: What would Bethe's life have been had he, like Peierls, obtained a job in a British university rather than at Cornell?

Rudolf Peierls, who died in 1995, was one year younger than Bethe. He grew up in Berlin in a somewhat similar familial and social background to Bethe's in Frankfurt. Both of Peierls's parents were assimilated German Jews; they had, in fact, converted to Christianity when the young Peierls was five years old. Bethe first met Peierls in Sommerfeld's seminar when Peierls came there in 1927. They became close friends, yet in their extensive correspondence until the early thirties they addressed each other as "Sie" rather than the more informal "Du." Both had to emigrate from Germany in 1933 and both went to England. Peierls remained in England when Bethe accepted his visiting appointment at Cornell in 1935. Peierls played a crucial role in establishing the feasibility of a U^{235} atomic bomb, and Bethe and Peierls were together at Los Alamos, Peierls being a group leader in charge of implosion calculations in the design of a plutonium bomb in the Theoretical Division that Bethe directed. Peierls returned to England in 1946, arguably the leading British expert on atomic energy and nuclear weapons. He was an occasional consultant at Harwell, the British atomic energy laboratory, giving technical advice on problems relating to isotope separation and the design of nuclear reactors. But the British political system is such that Peierls was never consulted on any policy matters relating to nuclear weapons. The British government relied on the upper echelons of the staff of Harwell and other governmental agencies, and on British-born scientists—e.g., John Cockcroft, James Chadwick, and William Penney—for advice on policy regarding nuclear weaponry, such as the building of an H-bomb or a moratorium on nuclear tests.[4]

There is little question that Bethe would have remained in Great Britain had he been offered a job there in 1934. If he had stayed there, it is clear that his life would have been very different. The scope of his political activities after World War II would surely have been narrower, and his potentialities in that sphere would probably never have been fully realized. Furthermore, I would argue that after World War II, Bethe's public activities as a citizen and his impact on nuclear policy as an adviser to the government were as, and perhaps more important than his intellectual productions as a scientist.[5]

In the biography, the chapter on Bethe and Edward Teller contrasts their reaction to the exercise of power; the one on Bethe and Victor Weisskopf addresses the esthetic dimension in their lives and work. The chapter on the moral responsibility of the scientist was intended to compare how Bethe and Oppenheimer confronted this issue, but in the course of writing the chapter grew to a size that made it unsuitable for inclusion in the biography. The writing of this chapter in its present form has not been without problems, for I know so much

Adam/
Lizzy

more about Bethe's life than about Oppenheimer's, and Bethe's story continues to unfold. Thus, finding a proper balance was difficult. I have tried to meet the challenge by concentrating on those events in which both played important roles, namely A-bombs and H-bombs. Nonetheless, an imbalance remains. Thus, chapter 6 deals almost exclusively with Oppenheimer's reflections on science and culture, because Bethe's are part of his biography.

In the Shadow of the Bomb is not a biography of either Bethe or Oppenheimer. It is principally concerned with the shaping of Bethe's and Oppenheimer's moral outlooks, how they constituted themselves as moral agents, how they assumed responsibility for their actions, and how they reflected on the Enlightenment and on modernity. Both in the original book and in the present one, I strove *not* to write a tragic account of Oppenheimer's life; nor do I want it to be seen as a contrast between good guy and bad guy.

Both Oppenheimer and Bethe were deeply affected by their Jewish roots. My extensive account of the Ethical Culture movement is meant to shed light on the sources of Oppenheimer's moral outlook and to give an insight into his coming to terms with being identified as a Jew, including the struggles this identification entailed. I do not recount some of the more familiar aspects of his life such as his "trial,"[6] nor do I describe an important facet of his later professional activities: his directorship of the Institute for Advanced Study. Rather, I focus on some of the compromises he made to acquire, secure, and maintain his influence and power. This is the reason, for example, for the detailed account of the Peters case. Similarly, in the case of Bethe, I do not address many of his accomplishments as a scientist and many of his activities in the public sphere. I am principally concerned with some of the factors that were responsible for his stature and authority as a public figure.

In the Shadow of the Bomb is also the story of the American physics community, although it is very much in the background. Physicists emerged from World War II as heroes—and theoretical physicists as *übermenschen*. It seemed—and at times these physicists believed—that they could do *anything*. In the public's eye, Oppenheimer became the embodiment of a new scientific persona: the scientist who had created new knowledge and new technologies that affected all of mankind, and who addressed the impact of these new technologies in both political and *moral* terms. But Oppenheimer was too fractured an individual to be able to carry the burden of that new persona. Bethe was more modest. He always strove to be integrated, always acted with integrity, and always knew where the anchor of his integrity lay. He became as concerned with the moral and the political as

Oppenheimer, and he came to embody the model of the responsible physicist. The point of my account of the Philip Morrison case is to illustrate that Bethe may not have been the most courageous nor the most vehement spokesman for academic freedom at Cornell or in the Newman Lab during the McCarthy era, but his actions and demeanor made explicitly clear to others what must be defended and what the right course of action was to be.

In the Shadow of the Bomb is an attempt to determine why Bethe became the embodiment of a new scientific persona: the scientist with exceptional technical expertise assuming the role of an intellectual. Like the intellectual of old, he addresses global problems affecting *all* of humanity, and he does so with moral concerns foremost in his mind.

Acknowledgments

A grant from the Alfred P. Sloan Foundation made possible the gathering of all the materials necessary to write Hans Bethe's biography. *In the Shadow of the Bomb* was written during 1997–98 while I was the recipient of a Dibner Fellowship. I thank Jed Buchwald, Evelyn Sihma, and the staff of the Dibner Institute at MIT for the warm hospitality that was extended to me there.

In writing this book I have had invaluable conversations with Monika and Rose Bethe. Similarly, I have benefited greatly from talks with Henry Bethe, Morton Camac, Dale Corson, Raine Daston, Wai Chee Dimock, Kurt Gottfried, Mary Haskell, Steve Heims, Hilde Hein, David Kayser, Evelyn Fox Keller, Jeri Lettvin, Henry Linschitz, Francis Low, Frank and Fritzie Manuel, Everett Mendelsohn, Philip Morrison, Melba Phillips, Barbara Rosenkrantz, Roger Smith, Ed Salpeter, Allen Sandler, Howard Schnitzer, Libby Schweber, George Stocking, Loup Verlet, and Charles Weiner. It is a better book for their comments and help. Cathryn Carson, David Cassidy, Freeman Dyson, Michael Gordin, Priscilla McMillan, and an unnamed reviewer for Princeton University Press carefully read a version of the manuscript and gave me very perceptive and valuable criticisms, which I have heeded while making revisions. I am very much in their debt. Also, I learned a great deal from reading Cathryn Carson's PhD dissertation on Heisenberg and Karl Hall's dissertation on Lev Landau and Soviet theoretical physics.

I would like to thank Elaine Engst, the archivist at the Kroch Library of Cornell University, in particular, and also the archivists at the Bancroft Library of the University of California/Berkeley, at the Institute for Advanced Study, at the Library of Congress, at the MIT Archives, at the National Academy of Sciences, at the Niels Bohr Library of the American Institute of Physics in College Park, Maryland, at the American Philosophical Society in Philadelphia, at the Los Alamos Archives, and at the University of Washington for making available materials in their archives and for permission to quote from these documents.

I have had the good fortune to have Alice Calaprice as the editor for both my *QED* book and the present volume. It is a gratifying experience to see one's manuscript being transformed into a lucid exposition, errors corrected, and bibliography and endnotes made accurate. She has performed these tasks expertly and gracefully. It is a real pleasure to work with her, and I greatly appreciate her assistance.

Despite all the help I have received, the responsibility for the content of the book, of course, rests with me. *Mea culpa*, for errors that are still present, for omission of references that might be relevant, or for inadvertently not giving appropriate credit for some of the materials.

The book owes much to my interactions with Paul Forman and Anne Harrington, two very special friends, whose insights I have freely appropriated. The book is dedicated to them, and to Everett Mendelsohn, a good friend and exceptional colleague, with affection and gratitude.

My greatest debt is to Hans Bethe.

IN THE SHADOW OF THE BOMB

INTRODUCTION

> Only strong personalities can endure history;
> the weak are extinguished by it.
>
> —Nietzsche

"Well, now we're all sons of bitches" was the trenchant comment made by Kenneth Bainbridge, the Harvard physicist who was in charge of the Trinity test, upon witnessing the explosion of the first atomic bomb at Alamogordo, New Mexico, at 5:29:45 A.M. on Monday, July 16, 1945, that fateful day that ushered in the atomic age.[1] The meaning of Trinity was obvious to Bainbridge at Alamogordo. But its moral and political implications were not as explicit nor as immediate to many of the other physicists there.[2] The "gravity" of the moral and political problems arising from the mastering of nuclear power had been carefully assessed by the physicists at the Metallurgical Laboratory (Met Lab) at the University of Chicago and forcefully spelled out in the Franck Report of June 11, 1945. That document was the memorandum drawn up by the subcommittee on Social and Political Implications of the Interim Committee that had been set up in May 1945 at the Met Lab to consider the implications of atomic energy. After estimating the casualties and damage that would result from the use of an atomic bomb, James Franck and his colleagues had concluded that

> the use of nuclear bombs for an early unannounced attack against Japan [was] inadvisable. If the United States were the first to release this new means of indiscriminate destruction upon mankind, she would . . . precipitate the race for armaments, and prejudice the possibility of reaching an international agreement on the future control of such weapons.
>
> Much more favorable conditions for the eventual achievement of such an agreement could be created if nuclear bombs were first revealed to the world by a demonstration in an appropriately selected uninhabited area.[3]

But that document had not been circulated among the Los Alamos scientists.[4]

While they had been aware of their singular contributions to the Allied victory during the war, only after the cessation of hostilities did

American physicists become fully conscious of the consequences of the bombing raids on Berlin, Dresden, Tokyo, among others; of the devastation of Hiroshima and Nagasaki; and of the role that scientists had played in these milestones. In his first major address after giving up the directorship of the Los Alamos project in the fall of 1945, J. Robert Oppenheimer, echoing the Franck Report, commented that

> atomic weapons were actually made by scientists, even . . . by scientists normally committed to the exploration of fairly recondite things. The speed of the development, the active and essential participation of men of science in the development, have no doubt contributed greatly to our awareness of the crisis that faces us, even to our sense of responsibility for its resolution.[5]

In the Shadow of the Bomb is about the crisis that Oppenheimer was referring to: the fact that after the war, American physicists had to confront the possibility that "the insight, the knowledge, the power of physical science, to the cultivation of which, the learning and teaching of which [they were] dedicated, [had] become too dangerous to be talked of."[6]

This book endeavors to sketch the self-consciousness, and the transformation, of American physicists during the beginning of the Cold War—the era spanning the Berlin blockade, the detonation of the first Russian atomic bomb (Joe 1), the Korean War, and the dawn of MADness (Mutual Assured Destruction)—by presenting some of their reflections on the novel roles they were assuming in the new world they had helped to create. It depicts their self-understanding by displaying some of their actions. That they were fashioning a new context had been apparent to them during the war, for they were being embedded in a new web of associations and were making warfare, heretofore almost principally a military activity, also into a civilian undertaking—and this on an unprecedented scale. With the beginning of the Cold War, many of them became convinced that they could not decouple themselves from this civilian-military affair. And all of them realized that after Hiroshima and Nagasaki they had to shoulder new responsibilities, for they had made untenable the notion that "civilians could leave the unpleasantness of war to the soldiers."[7] Their wartime inventions—radar, proximity fuses, atomic bombs—and the new gadgets they were creating after the war had made obsolete "the isolation of the world of science and of the intellect from that of politics and practical affairs."[8] Thus, in his foreword to the proceedings of the session on Nuclear Science at the Princeton Bicentennial Conference held in the fall of 1946, Eugene Wigner, one of the principal designers of the Chicago and Hanford nuclear reactors, noted that until very recently scientists for the most part had not participated in public life,[9] and when they had, "they did not serve the public *as scientists*." But it had become apparent that "the scientist *as scientist* will [now] have to face social responsibilities and human problems to an increasing degree."[10]

Their experience working at the Radiation Laboratory (Rad Lab) at MIT, at the Met Lab in Chicago, at Los Alamos, and at the other wartime laboratories had given American physicists a deceptive picture of their role in governmental affairs and of their standing in their partnership with the military. Vannevar Bush and James Conant had been given broad powers by Roosevelt, and the Office of Scientific Research and Development (OSRD), which they had headed, had put into *civilian* hands the responsibility for and control over the development of the weapons to be used by the army and the navy—functions and powers that had been the responsibility of the armed forces before the war.[11] The civilian scientists became the driving force in the partnership, and the alliance flourished during the war.[12] And with success came responsibility. Leo Szilard, one of the prime movers in the efforts to have the United States build an atomic bomb,[13] while working at the Met Lab in Chicago delineated the moral stand to be taken by the civilian scientists. In the fall of 1942, when the successful operation of the nuclear pile under Stagg Field became assured, Szilard felt that a stage had been reached where the physicists working on the bomb project had to choose between two options. One was to accept the hierarchical organizational structure and the compartmentalization that General Leslie Groves and the army wanted to impose on the project. This would relieve them of any responsibility for the construction and use of an atomic bomb since they would then serve as privileged hirelings taking their orders from higher-ups. Alternatively, they could commit themselves to the position "that those who have originated the work on this terrible weapon and those who have materially contributed to its development have, before God and the World, the duty to see to it that it should be ready to be used at the proper time and in the proper way."[14] For the most part, physicists opted for the second course of action. Although they saw to it that the bombs would be ready to be used "at the proper time," by the spring of 1945 it became clear that they would have little to say collectively about "the proper way" to use them.[15]

In the postwar period, the relationship between science and the military became shaped by the efforts of the armed forces to regain control over the planning and deployment of new weapons systems. But the posture of the American physicists still reflected the assumptions that underlay their wartime experiences: that the United States was a democracy; that the overall policy was set by civilians who had been elected by the citizenry at large; that they were the peers of the military personnel. And for the most part, physicists agreed with Oppenheimer that they had a special responsibility in helping resolve the problems posed by the new weaponry. They thus did not recognize until much later that physics—and science more generally—which heretofore in their lives as practitioners had been an *end*, was being transformed into a *means* for the state.[16]

The day-long session at the December 1946 annual conclave of the American Association for the Advancement of Science (AAAS) devoted to the topic "How far can scientific method determine the ends for which scientific discoveries are used?" is illustrative of the widespread concern of the scientific community about its new responsibilities. That AAAS meeting was held in Boston, and it was fitting for the organizers of the symposium to ask Percy Bridgman, the respected Harvard physicist and philosopher of science who had just been awarded the Nobel Prize, to present his views on these matters. Over the years he had forcefully made known his opposition to making science a servant of the state. He passionately believed that any restriction, any external imposition of an agenda, would corrode the purity of the scientific enterprise, and he had fiercely defended this position.[17] Bridgman made his lecture, to which he later gave the title "Scientists and Social Responsibility,"[18] a platform to challenge any social philosophy that required the *individual* scientist to be responsible for the use of his creations or the consequences of his discoveries. Bridgman did not believe that there was a scientific method as such, "but rather only the free and utmost use of intelligence,"[19] so he interpreted the theme of the symposium to be "What is the most intelligent way of dealing with the uses of scientific discoveries?" He took as his assignment the answering of the questions: "How far is it desirable that scientific discoveries be controlled?" and "What 'ought' to be the attitude of the scientist to his own discoveries?"

Bridgman was aware of the discussions that had taken place in Chicago in the spring of 1945 concerning the use of the bomb. He also knew about the speech that Oppenheimer had delivered to the Association of Los Alamos Scientists on November 2, 1945, in which he had counseled his fellow scientists to accept responsibility for the consequences of their work. As a result of discussions with his Harvard colleagues who had been at the Met Lab, at Los Alamos, and at the other wartime facilities, Bridgman had come to the conclusion that a large segment of the scientific community, "particularly those of the younger generation," and a large part of the public believed that scientists *are* responsible for the uses made of scientific discoveries. He interpreted this to mean that "each and every scientist has a moral obligation to see to it that the uses that society makes of scientific discoveries are beneficent." Should a scientist not meet that obligation, society could deem him culpable, and he could justifiably be disciplined with the loss of scientific freedom. But Bridgman strongly dissented from this view because he believed that it arose from a failure to realize both the larger and long-range implications of the relation of the individual to society. For Bridgman, the consummate individualist with hardy Puritan roots, it is the individual that is the entity that gives warrant to society: "Society is composed of you and me; society does not have an individuality of its own, but is the aggregate of what concerns you and me." The thesis of the responsibility of the individual scientist implied "the repudiation of the general ideals of the specialization and division of labor, and the ideal of, as far as possible, each man to his best." And this repudiation could only be justified if the assumption "that scientists are in some special way qualified to foresee the uses society will make of their discoveries, and to direct and control these uses" were true, and Bridgman did not accept this presumption. Nor did he believe that society had the right to exact "disproportionate service from special ability."[20]

Furthermore, Bridgman believed that society could deal with the issues raised by scientific discoveries by means other than forcing scientists to do something "uncongenial" for which they are not necessarily well suited. It was obvious to him "that if society would only abolish war, 99 percent of the control of scientific discoveries would vanish." He pointed out that since "the applications made of scientific discoveries are very seldom made by the scientists themselves . . . [i]t is the manufacture and the sale of the inventions that should be controlled rather than the act of invention." Moreover, this could be done by means at hand, such as revising the patent laws or forbidding Congress to fund scientific projects whose consequences are clearly deleterious. Some of the suggestions that Bridgman made in his lecture—in particular, that of banning injurious technologies—were later amplified by others when addressing the issues related to the manufacture of a hydrogen bomb.

Bridgman's position—that the individual scientist is responsible only to himself—might be branded hubristic, but in disclosing his apprehension he pointed out the generational conflict that had emerged:

> It is well known that the scientists who have shown the most articulate concern with the social concern of the atomic bomb are young. The philosophy that is coming into being betrays

this. It is a youthful philosophy, enthusiastic, idealistic, and colored by eagerness for self-sacrifice. It glories in accepting the responsibilities of science to society and refuses to countenance any concern of the scientist with his own interests, even if it can be demonstrated that these interests are also the interests of everyone.[21]

For Bridgman, the answers to the problems he had raised were to be found in education. He declared that the scientist's most important task was to make the average citizen recognize and feel "that the life of the intellect not only is a good life for those who actively lead it, but that it is also good for society as a whole that the intellectual life should be made possible for those capable of it, and that it should be prized and rewarded by the entire community." He concluded his lecture by stating that the most intelligent way to deal with the quandaries arising from scientific discoveries is to create an "appropriate society." And for him, that society was a scaled-up version of the scientific community:

> This society will be a society that recognizes that the only rational basis for its functions is to be sought in its relations to the individuals of which it is composed; a society in which the individual in his capacity as a member of society will have the integrity not to stoop to actions he would not permit himself as an individual; a society broadly tolerant and one which recognizes intellectual achievement as one of the chief glories of man; a society imaginative enough to see the high adventure in winning an understanding of the natural world about us, and a society which esteems the fear of its own intellect an ignoble thing.[22]

His vision of the "appropriate" society demanded that social, cultural, and intellectual diversity be tolerated. Most of the younger American physicists supported that position, though some might only have demanded that conduct be reasonable rather than rational. They were aware that the individual freedom demanded by scientific communities made these collectives models for all self-governing societies. The philosophical outlook of the post–World War I generations of American physicists had been deeply influenced by the pragmatism of Charles Saunders Peirce and William James. For Peirce and James, the *community* of science was a privileged one—indeed, a model for all democratic societies.

These physicists had become acquainted with pragmatism through the writings of John Dewey. Thus Dewey was extensively referred to in Edward Condon and Philip Morse's 1931 *Quantum Me-*

chanics, the first textbook on the subject published in the United States, one that was widely studied. In 1939, at a conference celebrating his eightieth birthday, Dewey gave a talk in which he incisively expounded his views on democracy as a moral ideal, convictions that had been stated, but not as concisely, in his earlier writings. For Dewey, democracy was a personal way of life to which citizens were committed in their everyday life:

> Democracy as compared with other ways of life is the sole way of living which believes wholeheartedly in the process of experience as end and as means; as that which is capable of generating the science which is the sole dependable authority for the direction of further experience and which releases emotions, needs, and desires so as to call into being the things that have not existed in the past. For every way of life that fails in its democracy limits the contacts, the exchanges, the communications, the interactions by which experience is steadied while it is enlarged and enriched. . . . [T]he task of democracy is forever that of creating a freer and more humane experience in which all share and to which all contribute.[23]

Democracy as a moral ideal as envisioned by Dewey might be difficult to achieve in the society at large, but the scientific community provided the proper conditions for its members to exercise intelligent judgment, to communicate in open exchanges, and to act "appropriately." During the 1930s, the American theoretical physics community could be said to aspire to Dewey's ideal.[24] Similarly in Germany, Max Weber, somewhat earlier, had made the post-1918 generation realize that science—though no longer the path to certainty nor a source for the meaning of the world—nonetheless provided, to those who chose it as a vocation, membership in a self-governing social body that offered them the possibility of "meeting the demands of the day" in human relations and the possibility of discharging their moral responsibility.[25]

The radical individualistic stand that Bridgman urged scientists to adopt was endorsed by but few other physicists. Bridgman himself recognized that gone were the days when an experimental physicist could do important and imaginative work by himself without adequate support to construct the necessary equipment and to purchase many of the instruments perfected during the war. Moreover, an ever larger segment of the fraternity depended on large and expensive pieces of equipment—cyclotrons, betatrons, Van der Graaf generators, cryostats, nuclear reactors—to carry out their research, equipment the government was willing to underwrite lavishly.[26] In addition, although not officially declared, the chills of the Cold War were beginning to be felt, and it was not clear whether physicists could or should isolate themselves in their ivory towers pursuing their own interests and setting their own agenda.

The temper of the times that Bridgman had alluded to in his AAAS speech was writ large at the Princeton bicentennial convocation. That conference is of particular interest because the lecturers were candid about their concerns, and the extensive discussions that followed each presentation were recorded and published.[27] Let me simply quote a few excerpts from some of the talks and record a few of the subsequent comments to indicate the atmospherics within the scientific community. In his presentation, Isidor Rabi worried about the enormous amounts of money available for the support of research because

> distributing large funds brings the distributors to the position in a short time of being able to apply certain pressures to the universities to change their method of working, to justify their activities, and conform their policies with other broad national and governmental activities. . . . [I]f it were decided to control universities and university research, there could be no better way to do this than the way it is being done now.[28]

In a comment after Rabi's talk, Harold Urey[29] suggested that it would be possible to avoid the government's complete domination by virtue of its lavish funding by the following strategy:

> If only we do not have unification of the armed services, if we have an army and navy which are separate and keep the tradional enmity between these two services, we may be in a position to escape this domination by playing one against the other. And if it should be possible to get a national science foundation established and to produce rivalry between the national science foundation and both the services, perhaps we should still be able to maintain our independence.[30]

Urey's statement in turn led George Kistiakowsky[31] to remark: "It seems to me that the danger is not in the government offering large sums of money to the colleges, but in the eagerness with which a great many scientists accept this money and sacrifice their freedom—accepting as they do, a specified program of research."[32]

The deliberations following the papers revealed the attempt by many of the participants to address the problem of responsibility by demarcating sharp boundaries between pure, basic, autonomous research—research that Vannevar Bush's influential July 1945 report on

Science: The Endless Frontier had characterized as "performed without thought of practical ends"—and applied, instrumental, utilitarian research which, according to Bush, was designed to provide answers "to important practical problems."[33] These discussions also indicated that the participants were deeply concerned with the issue that Wigner characterized as "the influences which the work of the scientist and the scientist himself exert or should exert on society."[34]

It was clear to the assembled physicists that the introduction of nuclear weapons required them to address *universal* problems, problems common to all humanity.[35] For as Oppenheimer had stated in his address in the fall of 1945 to the joint meeting of the National Academy of Sciences and the American Philosophical Society, the atomic bomb constituted "a vast new threat, and a new one, to all the people of the earth, by its novelty, its terror, [and] its strangely promethean quality."[36] The physicists' responsibility to address these universal problems stemmed from the fact that they had created the weapon, and their authority to do so devolved from their technical expertise. The issues raised by this expertise were addressed by the Princeton astronomer Henry Norris Russell, the dean of American astronomers, in a lecture delivered at a special session of the conference that was open to the public.[37] The question he posed was similar to the one Bridgman had addressed at the AAAS meeting, but his answer differed sharply from that given by Bridgman. He asked "should the scientific investigator be free, on his own recognizance, to experiment on any subject and any extent that he personally sees fit, and to publish his findings when and as he personally judges wise? Or should some sort of control be applied in certain areas? If so, in what cases, and why?"[38]

He answered these questions by pointing to the medical profession. Since physicians hold so much power over life and death in their hands, it is neither right nor safe to trust this power to their individual judgment. Hence, already in remote antiquity, they imposed upon themselves the Hippocratic oath to do no harm. Since physicists similarly now hold in their hands the power over the life and death of the species, Russell called for "a new Hippocratic code, for the physicist, and for all who deal with nuclear energy."[39]

His call went unanswered, but everyone connected with nuclear energy pondered the questions he had raised. Some—such as Philip Morrison, Robert Wilson, and Victor Weisskopf, who had been at Los Alamos during the war—foreswore working on weaponry of any kind. Many joined the ranks of the Federation of Atomic Scientists to have legislation passed that would place control over atomic energy in civilian hands and would establish effective international control over atomic weaponry to prevent an arms race.[40] That physicists played a special role in these efforts was natural, given their technical contributions during the war. In 1939, and especially after the fall of France in June 1940, many among them had felt it was their duty and calling "to save Western civilization."[41] The result was the formation of the Radiation Laboratory at MIT, the building of a nuclear pile at Chicago, and eventually the establishment of Los Alamos.[42] After the war, they felt it was their particular responsibility and mission to save mankind from the weapons they had invented.

What did it in fact mean for scientists to address problems affecting all of humankind? What moral and political responsibilities did it entail, particularly during the beginning of the Cold War and in the McCarthy era? And how did scientists respond to these demands? I have tried to answer these questions by looking at how two exceptionally gifted and influential men confronted their moral responsibility as scientists in the aftermath of World War II.[43] The two physicists are Hans Bethe and Robert Oppenheimer, whose lives became transmuted and intertwined by their wartime activities at Los Alamos. It is a brief account of how they dealt with the problems of trying to be responsible scientists and trustworthy citizens in the new world of antagonistic superpowers, each capable of annihilating the other with nuclear weapons.

By addressing these issues with a juxtaposition of the lives of these two men, it is possible to see more clearly how, among other things, character, culture, institutional context, and contingency helped shape particular stands taken. Bethe and Oppenheimer were members of the same generation, shared the same passion for science, and were transformed by their wartime experiences at Los Alamos. After World War II, they saw physics being recast from a vocation to a populous profession whose skills helped sustain the Cold War. They observed some of physics' most startling post–World War II innovations—the transistor and the laser—become incorporated, almost at once, into military technologies, and thus witnessed the further erosion of the "purity" of physics. Both became deeply involved in matters of national security and struggled to find effective means to reduce the danger posed by the ever-increasing number of nuclear weapons that were being developed and accumulated by the United States and the Soviet Union. All of these common experiences—including the fact that the families of both paid a heavy price for their professional activities and successes—makes a study in parallel lives meaningful.[44]

Both men were born in the first decade of the century: Oppenheimer in 1904, Bethe in 1906. Both grew up in secular homes and were educated in institutions whose ideals resonated with the culture of *Bildung* as envisioned by Wilhelm von Humboldt. For both of them, their Jewish descent was consequential. Both started their graduate studies with the advent of the new quantum mechanics and rode the crest of its successes. They came of age when physics had assumed a privileged position among the sciences. The importance of the physical sciences for utilitarian purposes had been recognized by the beginning of the century. The founding of institutions such as Nernst's Institute for Physical Chemistry and the Kaiser Wilhelm Institut in Germany, the initiation of the Solvay congresses and the inception of the Nobel prizes in physics and chemistry were all testimony to this. But physics, anchored in universal principles, came to be seen as being more "fundamental" than the other physical sciences. The transcendent qualities ascribed to it were in evidence in the worldwide acclaim Einstein received in the early 1920s upon the confirmation of his theory of general relativity. Albert Einstein, Max Planck, Hendrik Lorentz, and Niels Bohr gave theoretical physics a status unrivaled among the sciences.[45]

Bethe and Oppenheimer started their graduate studies when becoming a physicist was often a vocation, and theoretical physics was a discipline limited to a chosen few. By the mid-1920s, theorists differentiated themselves from their experimental colleagues by their mastery of mathematical tools and techniques and by the fact that their expertise usually encompassed *all* of physics. The microscopic world was certainly not "disenchanted" for them then,[46] for with the advent of quantum mechanics theoretical physicists were "representing" the lawfulness of that realm. In fact, they believed that quantum mechanics could teach them something about "the meaning of the world."[47] For Sommerfeld and his students, and Bethe in particular, this meant further verification of the "preestablished harmony between physics and mathematics" at the microscopic level.

Edward Teller has vividly described what it meant to enter the world of theoretical physics during the late 1920s: "I started my scientific work in Germany during the declining years of the Weimar Republic. For as long as I could remember, I had wanted to do one thing: to play with ideas and find out how the world is put together."[48] Theoretical physics was then a play form. Play—in contrast to work and leisure—is the rule-bound voluntary activity that is conducted within strict but arbitrary defined limits. It is pursued for its own sake, disinterestedly, with no material gain envisioned or intended. And theoretical physics as pure science and as play was an esthetic form without ethics.[49] But purity is an ideal form. It is based on a vision of conditions that need to be created and protected.[50] In the case of science it is a vision of intellectual order, one which entails a vision of social order.

Even though it was considered value neutral, one could aptly depict theoretical physics during the 1930s as a "religion which call[ed] for faith. [And] like every religion, it [had] its prophets, a college of apostles and the heart and soul of a whole people."[51] Furthermore, some of the sects were charismatic.

It would indeed be appropriate to describe Oppenheimer at Berkeley and at Los Alamos as having been bestowed with charisma, as having that "extraordinary quality of a person" by virtue of which he was "set apart from ordinary men and treated as endowed with . . . specifically exceptional qualities."[52] This extraordinariness manifested itself by the intensity with which he exhibited certain vital, crucial qualities: the breadth and depth of his erudition, the quickness of his mind, his ability to grasp almost immediately any material presented to him.[53] He was certainly deemed "charismatic" by his students at Berkeley during the 1930s and by the members of the staff of the Los Alamos Laboratory during the war. Max Weber's requirement that "this recognition [be] a matter of complete devotion arising of enthusiasm, or of despair and hope" was fulfilled.[54] Oppenheimer's students were passionate about physics. And at Los Alamos, despair stemmed from the causes of the war—national socialism, Hitler, Mussolini—and from the possibility that Germany might develop an atomic bomb before the Allies did, given that it had started on the project two years earlier than Great Britain and the United States. But there was also the hope that the gadgets Oppenheimer was having them build at Los Alamos would end the war and usher in a period of lasting peace.[55]

Weber's formulation tends to draw attention to the abnormal in the charismatic situation; but this need not be the case. Edward Shils has pointed to the distinctive character of the bond that fuses charismatic communities as a central and common element of both the extreme and the more routine expression of charisma:

> The charismatic quality of an individual as perceived by others,
> . . . lies in what is thought to be his connection with . . . some
> *very central* feature of man's existence and the cosmos in
> which he lives. The centrality, coupled with intensity, makes it
> extraordinary. . . . The centrality is constituted by its formative
> power in initiating, creating, governing, . . . maintaining . . .
> what is vital in man's life.[56]

The connection of theoretical physics with the *very central* features of the cosmos—as revealed by special and general relativity, and after

1925 by quantum mechanics—is what made it so distinctive and made the formulation of these theories such vital events. And in contrast to the *entzauberte* social and political world in which, as Max Weber had observed, belief in transcendent values and their embodiments in individuals and institutions was being driven into evermore restricted domains, theoretical physics after 1925 offered the possibility of engaging one's "heart and soul" in imposing order and coherence to the entire physical world. Moreover, one could do so by taking a pragmatic approach to the pursuit of knowledge and, in addition, preserve the integrity of contemplation.

During the 1930s, Oppenheimer embodied charismatic qualities both as a teacher and as an innovator and creator of representations that gave coherence to and provided understanding of physical phenomena. His students responded to these qualities with devotion and love. Teacher and pupils constituted a band of apostles that spread the gospel at Berkeley and Cal Tech—wherever they migrated.

The research activities of both Bethe and Oppenheimer during the 1930s, for the most part, were concerned with foundational problems far removed from possible applications.[57] They could then claim that science—certainly *their* science—was pure, and they could disavow responsibility for the uses made of the results of science. Their assertion that the ethical dimension only entered in the application of scientific knowledge assumed that boundaries could be drawn between basic, pure knowledge and applied, instrumental knowledge. It presumed that "fundamental," abstract, "objective" knowledge— "knowledge [obtained] as an end in itself, something to study because of the joy of it and the beauty of it"[58]—could be secured; and it assumed that this knowledge converged toward truth, though perhaps not toward a final truth. They shared the view that ethical foundations could not be sought in the material world. The decision on how to use science is a human decision, and it is in the social world that the ethical dimension enters.[59] Bethe on various occasions has stated that "such things as moral and aesthetic values are things that we make ourselves. No matter how strong the logic may be that forces us to accept certain moral values they are not moral values that are anywhere in nature. . . . Science has nothing to say or to contribute to these human values, for or against them. Nor can science tell us in which direction we should proceed to establish such values."[60]

World War II changed all that. Bethe is the supreme example why theoretical physicists proved to be so valuable in the war effort. It is his ability to translate his intellectual mastery of the microscopic world—that is, the world of nuclei, electrons, atoms, and molecules— into an understanding of the macroscopic properties of materials and *into the design of macroscopic devices* that rendered his services so valuable at Los Alamos.

The special role played by theoretical physicists during the war, especially at Los Alamos, is highlighted by the history of the atomic bomb. After the discovery of nuclear fission by Otto Hahn and Fritz Strassman in 1938 and the explanation of the phenomenon by Otto Frisch and Lise Meitner shortly thereafter, and the subsequent experimental evidence that more than one neutron is released in fission, the theorist Rudolf Peierls and Frisch formulated the initial conception of a fast neutron uranium bomb and made a rough estimate of the critical mass of U^{235} needed for such a device.[61] Its design was refined subsequently at Berkeley and at Los Alamos by theorists including Robert Serber, Hans Bethe, Richard Feynman, and Edward Teller. But the uranium bomb was never tested before its use over Japan—in contrast to the plutonium bomb with its complex implosion and ignition mechanism. Still, much of the success of the plutonium bomb derived from the extensive modeling and calculations performed by the Theoretical Division at Los Alamos. The theory/practice divide was irreversibly bridged there.

When physicists went back to their universities after the war, they tried to re-create the spirit of cooperation, commitment, and wholeness that had permeated the wartime laboratories, that of Los Alamos in particular. The Newman Laboratory for Nuclear Studies at Cornell was Bethe's attempt to do this, though it was to be a laboratory where only the "pure" science of subnuclear, high-energy physics, was to be carried out.

In his article on charisma, Shils noted that

> most humans beings, because their endowment is inferior or because they lack opportunities to develop the relevant capacities, do not attain that intensity of contact [possessed by the charismatic leader]. But most of those who are unable to attain it themselves are, at least intermittently, responsive to its manifestations in words, actions, and products of others who have done so. They are capable of such appreciation and occasionally feel a need for it. Through the culture they acquire and through their interaction with and perception of those more "closely connected" with the cosmically and socially central, their own weaker responsiveness is fortified and heightened.

This is surely part of the explanation of the response of the Los Alamos community to Oppenheimer's leadership. But Shils went on to say that all these "charismatic 'connections' may be manifested intensely in the qualities, words, action and product of individual personalities. . . . But they may also become resident, in varying degrees of intensity, in institutions—in the qualities, norms, and beliefs to which the members are expected to adhere or are expected to possess."[62] It is part of Bethe's distinction that he was able to endow the Newman Laboratory at Cornell with the "qualities, norms, and beliefs to which the members are expected to adhere or are expected to possess," qualities and norms that he was deeply committed to. In deep and meaningful ways, the identity of the members of the Newman Lab was defined to a large measure by this community of which they were so much a part. The laboratory created an environment that was a model of Dewey's communicative community: one that creates "a freer and more humane experience in which all share and to which all contribute," one that exists under the constraint of cooperation, trust, and truthfulness and is uncoerced in setting its goals and agenda; one that recognizes that human emancipatory interests are involved. Such communicative communities affirm that one of the most exalted of human aspirations—"to be a member of a society which is free but not anarchical"—can indeed be satisfied.[63]

After the war, the insistence to find boundaries between pure and applied research, between pure knowledge and technological knowledge, and the need to call Los Alamos an engineering project were all part of the process by which physicists came to terms with their wartime experiences. Pure knowledge was always a good. Applied research could be good or bad, depending on the goals and uses of research. Thus, in the fall of 1945 Oppenheimer emphasized the applied nature of the work at Los Alamos: "In the scientific studies which we had to carry out at Los Alamos, in the practical arts there developed, there was little of fundamental discovery, there was no great new insight into the nature of the physical world."[64]

Reconciling their sense of moral responsibility with their belief that "knowledge is a good in itself,"[65] even when that knowledge makes possible great evils, became an issue that physicists constantly had to face after the war. Oppenheimer put it thus in the fall of 1945:

> We have made a thing, a most terrible weapon, that altered
> abruptly and profoundly the nature of the world. We have
> made a thing that by all the standards of the world we grew up
> in is an evil thing. And by doing so, by our participation in making it possible, we have raised again the question of whether
> science is good for men, of whether it is good to learn about
> the world, to try to understand it, to try to control it, to help
> give to the world of men increased insight, increased power.

And the answer he gave to the question was to make explicit his commitment to the Baconian vision of science: "Because we are scientists, we must say an unalterable *yes* to these questions: it is our faith and our commitment, seldom made explicit, even more seldom challenged, that knowledge is a good in itself, knowledge and such power as must come with it."[66]

Somewhat later he asserted:

> When you come right down to it the reason that we did this job [building the atomic bomb] is because it was an organic necessity. If you are a scientist you cannot stop such a thing. If you are a scientist you believe that it is good to find out how the world works; that it is good to find out what the realities are; that it is good to turn over to mankind at large the greatest possible power to control the world and to deal with it according to its light and its values. . . . It is not possible to be a scientist unless you believe that it is good to learn . . . unless you think that it is of the highest value to share your knowledge, to share it with everyone that is interested. It is not possible to be a scientist unless you believe that the knowledge of the world, and the power that this gives, is a thing which is of intrinsic value to humanity, and that you are using it to help in the spread of knowledge, and are willing to take the consequences.[67]

In addition, in his November 2, 1945, speech at Los Alamos, Oppenheimer had declared that "secrecy strikes at the very root of what science is, and what it is for."

Like Oppenheimer, Edward Teller, one of the most vociferous advocates of building the Super, the H-bomb, believed that *all* scientific knowledge is good and that this knowledge must be open, for secrecy is antithetical to its growth. In fact, Teller concluded his article in *Science* in 1955 on the history of the development of the Super with a statement not very different from Oppenheimer's:

> We would be unfaithful to the tradition of Western civilization if we were to shy away from exploring the limits of human achievement. It is our specific duty as scientists to explore and to explain. Beyond that our responsibilities cannot be any greater than those of any other citizen of our democratic society. . . . I am confident, whatever the scientists are able to discover or invent, that the people will be good enough and wise enough to control it for the ultimate benefit of everyone.[68]

And in an interview in 1994 in which he recalled the H-bomb debate, Teller went even further: "There is no case where ignorance should be preferred to knowledge—especially if the knowledge is terrible."

But it was Feynman who best expressed the views of Oppenheimer, of Teller, and those of many of the other scientists who had been at Los Alamos. He first stated his position in a public address at the 1955 autumn meeting of the National Academy of Sciences held on the Cal Tech campus[69] and again in the first of the lectures he delivered at the University of Washington in 1963. Echoing Oppenheimer, Feynman put the matter thus:

> I think a power to do something is of value. Whether the result is a good thing or a bad thing depends on how it is used, but the power is a value.
>
> Once in Hawaii I was taken to see a Buddhist temple. In the temple a man said, "I am going to tell you something that you will never forget." And then he said "To every man is given the key to the gates of heaven. The same key opens the gate of hell."
>
> And so it is with science. In a way it is a key to the gates of heaven, and the same key opens the gate of hell, and we do not have any instructions as to which is which gate. Shall we throw away the key and never have a way to enter the gates of heaven? Or shall we struggle with the problem of which is the best way to use the key? . . . I think we cannot deny the value of the key to the gates of heaven.[70]

The parable is indeed unforgettable—but it is not clear that its applicability in the case of individual action justifies its extrapolation to the collective activity Feynman calls science. Who is the "we" Feynman is referring to? Who are the "we" who decide whether to throw the key away or to struggle?

Bethe, Feynman, Oppenheimer, and Teller all regarded scientific knowledge as an enabling power to do either good or evil. They would all agree with Nicholas Rescher's assessment in his *Forbidden Knowledge* that there may well be some things we ought not to know. There might exist "some information [that] is simply not safe for us—not because there is something wrong with its possession in the abstract, but because it is the sort of thing we humans are not well suited to cope with." But to the question "Are there also moral limits to the *possession* of information per se—are there things we ought not to know on moral grounds?" Rescher gives the answer: "Here inappropriateness lies only in the mode of acquisition or in the prospect of misuse. *With information, possession in and of itself—independently of the matter of acquisition and utilization—cannot invoke moral impropriety.*"[71]

Pure science cannot invoke moral impropriety. Although after 1945 it became ever more difficult to draw sharp boundaries between pure and applied science,[72] between applied science and technology, and to insulate "pure," basic science from both its consequences and its requirement of societal support,[73] Oppenheimer in 1947 would still try to make a distinction based on motivation, just as Vannevar Bush had done earlier in *Science: The Endless Frontier*.[74] But he had come to accept the fact that the boundaries had become ambiguous and permeable. He, of course, recognized the debt of science to technology—that technology often drives science—and he was aware of the mutual interconnection, interdependence, and cross-fertilization of science and technology. He was also certainly conscious of the fact that science could spawn instruments of destruction. Nonetheless, Oppenheimer indicated that while he would not argue against the notion of scientists assuming responsibility for the fruits of their work, the realities were such that "it must be clear to all of us how very modest such assumption of responsibility can be, how very ineffective it has been in the past, how very ineffective it will surely be in the future." For him, "The true responsibility of the scientist . . . is to the integrity and vigor of his science."[75]

The issue of the responsibility of scientists for the knowledge they create became central in the debate over the development of hydrogen bombs, and later in the justification of their manufacture. Perhaps the wartime conditions and the evident evils of Hitler's totalitarian Nazi state justified the development of atomic bombs. However, given that the United States was in possession of atomic weapons, given the magnitude of the devastation a hydrogen bomb would wreak, should one participate in the efforts to build this bomb? Do the reality of the Cold War and the nature of Stalin's totalitarian regime trump these moral scruples? In the fall of 1949, Oppenheimer, concerned with the dangers of acquiring some forms of technological knowledge, *though in favor of exploring the feasibility of a fusion bomb*, argued against producing the hydrogen bomb and thereafter sought ways to ban the technologies for building such weapons.

Oppenheimer's concerns regarding the danger of the knowledge of certain technologies have persisted. They have, in fact, become more acute, for simulations and "virtual" testing using modern, powerful high-speed computing and advanced graphics have almost eliminated the need for "real" testing. Should some knowledge be forbidden? If yes, on what ground? How is the kind of knowledge that might be forbidden characterized? Should some technologies be forbidden? These questions have become particularly consequential issues for the evolution of our species, for with the advances in genetics and developmental biology we are acquiring the power to directly shape our own evolution.

The issues involved when posing the question, Should some knowledge be forbidden?, are complex. And it should be clear that the answers to whether some technologies should be forbidden are context, cost, and scale dependent.[76] Furthermore, we must remember, as Freeman Dyson and others have stressed, that the assertion that technology drives and shapes ethics has become as true and as compelling as Max Weber's claim that ethics drives technology.[77]

Interestingly, Bethe has maintained the distinction among pure science, applied science, and technology. It is not that he is unaware of their interconnections and mutually fructifying interactions. Rather, his persistence stems from his continuous involvement in all three areas since World War II and from the fact that he is able to compartmentalize these activities in his own professional life.

The distinction between pure and applied science, particularly in the nuclear realm with its connection to atomic bombs, was of course blurred in the public's mind after World War II. The population at large had in fact been completely unprepared for the atomic bomb, unaware of its revolutionary impact on war technology, and unaware of the desperate need to avoid any major war.[78] But it learned quickly, partly as the result of the activities of the Federation of Atomic Scientists. Physicists—whether engaged in "pure" or "applied" science—came to be seen as requiring conscience by virtue of their activities and their expertise. Oppenheimer accepted this demand and made this explicit in a famous address at MIT in 1947. He there stated that the threat of Nazi Germany had made physicists become "reluctantly . . . aware of [their] dependence on things which lie outside science," and that despite the vision and the far-seeing wisdom of Churchill and Roosevelt,

> the physicists felt a peculiarly intimate responsibility for suggesting, for supporting, and in the end, in large measure, for achieving the realization of atomic weapons. Nor can we forget that these weapons, as they were in fact used, dramatized so mercilessly the inhumanity and evil of modern war. In some sort of crude sense which no vulgarity, no humor, no over-statement can quite extinguish, the physicists have known sin; and this is a knowledge they cannot lose.[79]

His statement summed up the new moral and political dilemmas of the age. As Don Price observed in his eulogy for Oppenheimer,

> The new powers that science had conceived and engineering had delivered had destroyed the innocence and the sense of freedom of the scientist. Henceforth the scientist could never

profess a lack of responsibility for the fate of society; yet whenever he responded to the call to political action, he would have to deal with problems that far transcended his specialized knowledge.[80]

Nonetheless, as Percy Bridgman had emphasized, scientists had a duty to educate the citizenry so that they could comprehend the new world they were living in and be able to make informed decisions about its future. Both Oppenheimer and Bethe assumed these new responsibilities with zeal and dedication. They became deeply involved in political activities, both within the government and before the public. Oppenheimer became an official adviser to the State Department and to the armed forces, and until 1954—when the Atomic Energy Commission (AEC) revoked his clearance—he was one of the most influential civilians shaping American policy regarding nuclear power and nuclear weapons. This was especially so after the creation of the Atomic Energy Commission in 1946, when he became the chair of its General Advisory Committee (GAC).[81] He was an eagerly sought after public speaker,[82] and he used these occasions to alert the nation to the need to think boldly and creatively regarding nuclear energy. Bethe, too, became politically active: he participated in the postwar efforts of the Federation of Atomic Scientists to guarantee civilian control over atomic energy;[83] he became a charter member of the Emergency Committee of Atomic Scientists, testified extensively before congressional committees, and gave frequent public lectures on the perils and hopes of atomic energy. But until 1957, when he became a member of PSAC (President's Science Advisory Committee), the major part of his activities outside Cornell was devoted to giving technical advice to Los Alamos and to industrial firms seeking to develop nuclear power. To this day, Bethe believes that the benefits that will accrue from the peaceful uses of atomic energy will justify the unraveling of nuclear power.

In important ways, Bethe and Oppenheimer gave different answers to the question, What is the role of the scientist in a democracy?[84] They agreed that, after Hiroshima and Nagasaki, scientists had a special contribution to make by virtue of their technical expertise and had an obligation to become involved in the political process. Oppenheimer chose to assume the role of the insider, exerting great influence as a highly admired adviser within the government. This gave him enormous power but also entailed compromises.[85] Bethe took a more independent stand. He became a trusted technical adviser to the government, maintaining close ties with Los Alamos; but he assumed an outsider's critical stance when he felt that either moral scruples or technical considerations required a different position than that "officially" advocated.[86]

The contrast between Bethe and Oppenheimer in the political arena had a parallel within physics. Oppenheimer not only adopted the new role of scientist-statesman in American political life and became a national public figure, he also became one of the senior statesmen of American physics after accepting the directorship of the Institute for Advanced Study in 1947.[87] Thereafter, he no longer did any original research in physics and became primarily concerned with synoptic assessments.[88] He came to believe that if a few wise people at the top had the right ideas, they could effect change. In both politics and physics, he was a member of that elite and perhaps assumed that therefore things would be fine. For a time, he may in fact have believed that only he had the pertinent insights and appropriate answers in matters of nuclear policy, and that only his participation could bring about a safe nuclear world. Perhaps his singular role at Los Alamos made him feel that he had a unique responsibility for finding solutions to the threat nuclear energy and nuclear weapons posed for humankind.

Bethe, by contrast, has been more modest. He has constantly remained the working craftsman, always engrossed in mastering all the details and always himself carrying out all the nitty-gritty aspects of whatever technical problem he has been involved with. He has acted on his belief that the most effective way for him to exercise responsibility is by being an active and productive scientist. He recognized that, by providing his skills and expertise to the government, he also gained access to the decision-making process. He used this entry to influence policy directly as when he served on PSAC and its subcommittees. In situations where he disagreed with the government's position, as in the hydrogen bomb decision, he criticized these policies in other forums, and his criticisms were influential because of the knowledge he had acquired as an active participant in the technical research process.

Mastery of physics has been and remains Bethe's anchor in integrity. The role of elder statesman in both science and politics has come to him by virtue of the respect and esteem the physics community and the nation have for him, and not because he sought that role. The letter sent to Bethe by President Lyndon Johnson on the occasion of his sixtieth birthday summarizes why he became a national treasure:

> You are not only an outstanding scientist, you are also a devoted public servant.

The nation has asked for your help many times and you have responded selflessly. You have made profound contributions in the fields of atomic energy, arms control and military technology. And you have been an important source of the immense contribution which science and the university community are making to society as a whole.

Our country is deeply indebted to you.[89]

Perhaps the different fates that Bethe and Oppenheimer wrought for themselves can be attributed to what Nietzsche had called a universal law:

A living thing can only be healthy, strong, and productive within a certain horizon; if it is incapable of drawing one round itself, or too selfish to loose its own view in another's, it will come to an untimely end. Cheerfulness, a good conscience, belief in the future, the joyful deed—all depend, in the individual as well as the nation, on there being a line that divides the visible and clear from the vague and shadowy; we must know the right time to forget as well as the right time to remember, and instinctively see when it is necessary to feel historically and when unhistorically.[90]

Bethe knew this instinctively. It was something Oppenheimer grasped, but with great difficulty.

For all their differences in style and approaches to physics and politics, and though they were born in different countries and shaped in different environments, Oppenheimer and Bethe shared a common vision: the legacy of the Enlightenment. Both believed that the physical universe was comprehensible. They both had faith in Reason, both believed that knowledge is good, that scientific knowledge is good and apolitical, that it should be open and shared, and that it will lead to progress.[91] Thus, in 1953, echoing the French philosophe, the Marquis de Condorcet, Oppenheimer would assert that because of science, "this is a time that tends to believe in progress. Our ways of thought, our ways of arranging our personal lives, our political forms, point to the future, point not merely to change, to decay, to alteration, but point with a hopeful note of improvement that our progress is inevitable."[92]

But note that this was a somewhat tempered championing of Enlightenment ideals. Oppenheimer was restrained: "It is a time that *tends* to believe in progress," and he *hopes* that "progress is inevitable." Condorcet's prophecy that "the time will come when the sun will shine only on free men who have no master but their reason"[93] had lost some of its luster. Already in 1939, Dewey, upon noting the emergence of totalitarian states in Germany and in the Soviet Union, had written: "It is no longer possible to hold the simple faith of the Enlightenment that assured advance of science will produce free institutions by dispelling ignorance and superstition." Oppenheimer had likewise given up this "simple faith" when Germany and the Soviet Union signed a nonaggression pact just before the outbreak of World War II. But he had not given up his faith in science and reason.

Bethe and Oppenheimer were aware that an age or a culture is characterized by the extent of its knowledge and by the nature of the questions it asks. The extent had become infinite; and given human finitude, the answers to the questions posed could only be tentative. They recognized that science, morality, and esthetics had become severed into autonomous spheres; that the problems that had devolved from the unified Weltanschauung of revealed religion and metaphysics of an earlier Christian age had been recast as problems of validation: What is true? What is right? What is beautiful? And with the reordering, these questions had become addressed as problems in epistemology, in morality and justice, and in taste. Discourses about knowledge, theories of morality and jurisprudence, questions regarding the production and criticism of art had become the concerns of institutionalized cultural professions and were dealt with by specialized experts. As a result, a chasm had developed between the culture of the various experts, and an even greater gulf developed between these cultures and the culture at large.[94] The problem of grounding moral action and social bonds in this fractured world—in which religion and metaphysics had lost their authority—was faced by both of them. Both Bethe and Oppenheimer strove to bridge the divides. Both believed that knowledge contributes to the good of mankind because knowledge allows men and women to make rational choices and thus bolsters their freedom: the growth of knowledge is therefore good in itself. Both had been brought up on Kantian ideals and believed that the faculty of reason enables one to arrive at answers to how life is to be lived and what is to be done in all spheres: the moral, the social, the political, the practical. Although they thought that the answers that reason provided are valid for all rational human beings under the same conditions, they recognized that it would be difficult to obtain universal assent as to what constitutes being "rational." But "rationality" within science seemed unambiguous. Science for them became a faith practiced by an international fraternity,[95] a faith that essentially staked everything on the power of reason to render the world comprehensible and to mold it to human ends. They recognized that "science is not all of the life of reason," but they had confidence—and were encouraged by looking at past history—that "science, as one of the forms of reason, will nourish all its forms."[96]

However, after World War II they had to confront the fact that the seeming triumph of rationality had made reason become the scaffolding of the irrational. Auschwitz, Bergen-Belsen, Birkenau, Dachau—the murder of millions of Jews, Gypsies, and others—had undermined the idea of progress in history.[97] Similarly, the unraveling of the structure of the atom and of the nucleus had created a world in which the escalation of military power, rather than increasing security, had proliferated new dangers, a world whose annihilation was being averted by a balance of terror.[98] And the political environment in Soviet Russia had made them address the question whether tolerance and compromise were the way to deal with those who have accepted a totalitarian vision.

The confrontation reached a critical point during the Vietnam War. During a sit-in at MIT on March 4, 1969, Victor Weisskopf succinctly and movingly stated the beliefs that had motivated Bethe's, Oppenheimer's, and his own actions:

> We scientists are optimists. We believe that rational thought and planning will be able to rectify the ills that technology has caused. We believe that, at the end, much good will come from the applications of science. But science does not only influence our physical environment, it also creates our mental environment. It has deep influence on our thinking and our outlook; it is an integral part of our civilization. It is an activity in which our modern culture has been most creative and successful. Therefore, we must not neglect, even in these days of crisis, our responsibility to science itself, to the continuation of our great search for truth and meaning in the material world, our quest to know more about the universe in which we live. It is this quest that has brought nations and continents together, it is this quest whose ideas and ideals transgress national and political borders. It is a human quest because it is the cause of humanity as a whole. Whatever your viewpoint, it is good to know more.[99]

This book can be read as an account of the way Bethe and Oppenheimer each dealt with the inadequacies of the Enlightenment ideals. Before World War II, Berkeley and Cornell—and during the war, Los Alamos—had given Bethe and Oppenheimer the solidarity that close-knit communities with a common vision and purpose give to their members. After the war, they came to regard comradeship and communion with their fellow human beings evermore important facets of their lives. Thus Oppenheimer, in his last address at Los Alamos, noted:

To these deepest things which we cherish [such as democracy], and for which Americans have been willing to die . . . even in these deepest things, we realize that there is something more profound than that; namely, the common bond with other men everywhere. . . . We cannot forget our dependence on our fellow men; . . . also our deep moral dependence, in that the value of science must lie in the world of men, that all our roots lie there. These are the strongest bonds in the world, . . . these are the deepest bonds—that bind us to our fellow men.[100]

Both Bethe and Oppenheimer came to see fellowship and a sense of common purpose as two of the profoundest and most resonant of human values. The redemptive quality of the affection and love that accompanied that communion became a prized and cherished good. But whereas Oppenheimer—perhaps because of his temperament and the nature of the Institute for Advanced Study—saw salvation in individual terms, Bethe sought it at the level of community. He helped build a remarkable intellectual and social community at Cornell that gave him sustenance, fortitude, and *caritas*.

In the Shadow of the Bomb narrates the shaping of Bethe's and Oppenheimer's moral outlook. Because rationality and Enlightenment ideals are so central to both Bethe's and Oppenheimer's core beliefs, in chapter 1, I review Kant's analysis of "What is Enlightenment?" as presented in Foucault's exegesis of it. Chapter 2 depicts the life of the young Oppenheimer, with an emphasis on Felix Adler and the Ethical Culture movement. It is my contention that, despite important derelictions, Oppenheimer was a deeply moral person, and that some of his actions after Los Alamos can only be understood as attempts at expiation and atonement. How else can we explain his coming to President Harry Truman in 1946 and exclaiming: "I have blood on my hands"?[101] But it is a cri-de-coeur expressed in Christian evocations alluding to original sin: "In some sort of crude sense which no vulgarity, no humor, no over-statement can quite extinguish, the physicists have known sin; and this is a knowledge they cannot lose."[102] Oppenheimer's moral outlook was shaped, in part, by his attending the Ethical Culture School, and chapter 2 is an exploration of that culture. In chapter 3, I briefly sketch Bethe's biography, and some of Bethe and Oppenheimer's interactions. In chapter 4, I review some of their activities during the McCarthy era in cases involving colleagues, and in chapter 5, I discuss the position they took on nuclear weapons. In chapter 6, I cover some of their views on the relation between science and society and between the individual and society. An epilogue attempts to bring the various themes together.

1.

WHAT IS ENLIGHTENMENT?

> Enlightenment is man's leaving his self-caused immaturity. Immaturity is the incapacity to use one's intelligence without the guidance of another. Such immaturity is self-caused if it is not caused by lack of intelligence, but by lack of determination and courage to use one's intelligence without being guided by another. *Sapere Aude!* Have the courage to use your own intelligence! is therefore the [heraldic] motto [*Wahlspruch*] of the enlightenment.
>
> —Immanuel Kant (1949, 132)

In November 1784, the *Berlinische Monatschrifte* published Kant's response to the question, "What is Enlightenment?" which the magazine had posed earlier that year. Kant's now well-known answer was given in the epigraph above. Some two hundred years later, Michel Foucault took Kant's essay as the point of departure for his reexamination of this same question: *Was ist Aufklärung?*[1] The importance of Kant's essay for Foucault stemmed from the fact that he saw it as a watershed: "modern" philosophy could be characterized as the philosophy that is attempting to answer the same question as the *Berlinische Monatschrifte* had raised.[2] Foucault suggested that for Kant the importance of his "little text" derived from the fact that it gave him the opportunity to assess the contemporary status of both his own philosophic enterprise and his reflections on history, and to examine how these two activities intersected.[3] And by looking at Kant's essay in this way, Foucault proposed to connect Kant's *Aufklärung*, the leaving of immaturity, with what he called the "attitude of modernity" with its consciousness of contemporaneity, "a modernity which sees itself condemned to creating its self-awareness and its norms out of itself."[4] *Creating the norms that were to guide them as moral agents in* and *out of* the new world they had helped create were central problems Bethe and Oppenheimer addressed after the war; and they addressed them as children of the Enlightenment.

This chapter draws on Foucault's incisive article, so it may be helpful to summarize it briefly. Foucault stressed that Kant had defined

28

two essential conditions under which mankind can escape from its immaturity. The first is that the realm of obedience and the realm of reason must be clearly distinguished. Humanity will reach maturity when it is no longer required to obey any authority that demands "Don't think, just follow orders" and when men are told, "Obey, and you will be able to reason as you like." But Kant distinguished between the public and the private uses of reason: "Reason must be free in its public use, and must be submissive in its private use. Which is . . . the opposite of what is ordinarily called freedom of conscience."[5] Man makes private use of reason when he is "a cog in a machine," that is, when he has a role to play in society and jobs to do. To be a soldier, to have taxes to pay, to be a civil servant were the examples that Kant had given. Under those circumstances, man finds himself placed in a circumscribed situation, where he has to apply particular rules and pursue particular ends. In these situations his reason must be subjected to the particular ends in view, so that there cannot be any free use of reason. But when one is reasoning only to use one's reason, when one is reasoning as a reasonable human being as a member of reasonable humanity (which is the meaning of the German word *räsonieren* as Kant used it), then the use of reason must be free and public. "There is Enlightenment when the universal, the free and the public uses of freedom are superimposed on one another."[6] Enlightenment must thus not be conceived simply as a general process affecting all humanity, nor as an obligation prescribed to individuals. It also poses a political problem: How can the use of reason take the public form that it requires? How can *Sapere Aude!* be exercised publicly, while individuals are obeying scrupulously privately? (This, incidentally, was precisely the problem Bethe and Oppenheimer faced in connection with the H-bomb!) The solution for Kant was to propose a sort of contract to Frederick II—the contract of rational despotism with free reason: "The public and free use of autonomous reason will be the best guarantee of obedience, on condition, however, that the political principle that must be obeyed itself be in conformity with universal reason."[7]

At this point Foucault left Kant's "brief article" and turned to its connection to Kant's three *Critiques* and to the linkage of *Aufklärung* to modernity. Foucault conceived modernity as an "attitude," "a mode of relating to contemporary reality; a voluntary choice made by certain people; in the end, a way of thinking and feeling; a way, too, of acting and behaving that at one and the same time marks a relation of belonging and presents itself as a task. A bit, no doubt, like what the Greeks called an *ethos*."[8] By stressing choices and behavior, Foucault again made clear that the task at hand was principally a moral one.

To make his conception of modernity more concise, Foucault pointed to its characterization as a consciousness of the discontinuity of time: "a break with tradition, a feeling of novelty, or vertigo in the face of the passing moment," and he quoted Baudelaire's definition of modernity: "the ephemeral, the fleeting, the contingent, the half of art whose other half is the eternal and the immutable."[9] Foucault commented that, for Baudelaire,

> being modern does not lie in accepting this perpetual move-
> ment; on the contrary, it lies in adopting a certain attitude with
> respect to this movement; and this deliberate, difficult attitude
> consists in recapturing something eternal that is not beyond
> the present instant, nor behind it, but within it. . . ; modernity
> is the attitude that makes it possible to grasp the "heroic" as-
> pect of the present moment. Modernity is not a phenomenon
> of sensitivity to the fleeting present; it is the will to "heroize"
> the present.[10]

For Baudelaire, modernity was a mode of relationship that has to be established with oneself. Modern man dedicates himself to asceti-cism and commits himself to a discipline whereby he does not go off to discover himself, but instead tries to invent himself. "Modernity does not 'liberate man in his own being'; it compels him to face the task of producing himself. . . . But Baudelaire did not imagine that this heroization of the present and ascetic elaboration of the self had any place in society itself, or in the body politic. They can only be produced in another, a different place, which Baudelaire calls art."[11]

Toward the end of his article, Foucault stressed that it was not im-portant whether he had summarized successfully the complex histor-ical event that was the Enlightenment, or depicted effectively the atti-tude of modernity in the various guises it may have taken during the last two centuries. Nor is it important for me whether his exposition and interpretation of the Kantian canon meet the approval of Kant scholars. I find attractive his suggestion that the thread that may con-nect us with the Enlightenment is not a faithfulness to doctrinal ele-ments, but rather a permanent reactivation of an attitude—that is, of a philosophical ethos that could be described as a permanent critique of our historical era whose aim "will be oriented toward the 'contem-porary limits of the necessary,' that is, toward what is not or no longer indispensable for the constitution of ourselves as autonomous sub-jects."[12] And here Foucault clearly rejected Kant's claims of essential-istic a priori limitations intrinsic to our very constitution as thinking and willing subjects and Kant's view of ethics as fixed and transcen-dent in some way.

In concluding his article, Foucault indicated that what is at stake are the answers to the following question: "How can the growth of the capabilities of individuals with respect to one another be severed from the intensifications of power relations that are conveyed by various technologies (for example, institutions whose goal is social regulation, or productions with economic aims, or techniques of communication)."[13] To answer the question would lead to the study of what Foucault called "practical systems," by which he meant

> what [people] do and the way they do it. That is, the form of rationality that organizes their ways of doing things (this might be called the technological aspect) and the freedom with which they act within these practical systems, reacting to what others do, modifying the rules of the game, up to a certain point. . . . [The study of these practical systems] will have to address the questions systematized as follows: How are we constituted as subjects of our own knowledge? How are we constituted as subjects who exercise or submit to power relations? *How are we constituted as moral subjects of our own actions?*[14]

This is Foucault's translation of Kant's famous threefold question: What can I know? What should I will? and, What may I reasonably hope for? And the central question is a moral one.

Rejecting Kant's universalistic response, Foucault believes the legacy of Kant's reflection is that *Aufklärung* has to be considered not "as a theory, a doctrine, nor even as a permanent body of knowledge that is accumulating; it has to be conceived as an attitude, an ethos, a philosophical life in which the critique of what we are is at one and the same time the historical analysis of the limits that are imposed on us and an experiment with the possibility of going beyond them."[15]

I would like to suggest that it is in this Foucaultian sense that both Bethe and Oppenheimer are children of the Enlightenment. Both had been raised on Kant's universalist maxims on morality—Bethe in the *Gymnasium*, Oppenheimer in the Ethical Culture School where Felix Adler's emendation of them were guiding tenets. Both Bethe and Oppenheimer had sought certain universal values.[16] Both became very much concerned with the self-shaping, volitional aspect of ethical conduct; both came to include contextual factors and culture-specific values and motives in making sense of themselves as moral agents. Thus, in a lecture delivered at the University of North Carolina in 1960, Oppenheimer asserted: "It was one thing to say, along the banks of the Sea of Galilee, 'Love thy neighbor.' It is a different thing to say it in today's world. Not that it is less 'true'; but it has a different meaning in terms of practice and in terms of what men can manage."[17]

Oppenheimer could be called a relativist, for he was not sure that one could recognize a detached or valid perspective from which to judge the morality of other societies. In a letter in 1951 to George Kennan, who found it difficult to understand his sense of morality, Oppenheimer explained:

> It is not in our judgment of ourselves or our own actions that I would reject moralism: it is rather in our attitude toward the behavior of other peoples. What I question is our ability to put ourselves, as a nation, in the place of these other peoples and decide what is right or wrong in the light of their standards and traditions, as they see them, or even in the eyes of the Almighty. I regard the behavior of other societies as something the morality of which I would prefer not to have to determine. I think it is our business to study that behavior attentively, to measure the intensity of the emotional forces behind it, and to take careful account of the potency of its influence on international affairs; but I feel we would do better not to attempt to classify it as "right" or "wrong", praiseworthy or reprehensible. We Americans have enough, it seems to me, with our consciences and with the necessity, now upon us, to reconcile an individualistic tradition with the centralizing pressures of advanced technology. It is for this that we are accountable as a body politic, not for the decisions and solutions arrived by others. Let us conduct our policies in such a way that they are in keeping with our own character and tradition. This means, of course, that the moral element, as we feel it, must be present.[18]

Both men came to lead a life of personal inquiry in which the examination of who and what they were was at the same time an analysis of the limits that were imposed on them and an experiment to determine the possibility of going beyond them.

Both Bethe and Oppenheimer were conscious of the strong, moral influence of the social dimension of their scientific activities. They staunchly believed that, as scientists engaged in fundamental physics, they had assumed a privileged role—and a special responsibility—as members of a community can be described as committed to the Peircian vision; that is, a community committed to rationality—but not instrumental rationality—for which communication inheres in its very being, whose members believe in a basic ontology of the world and affirm that it is possible to decipher and ascribe order to the physical universe.[19]

On Foucault's analysis, Bethe has remained more Kantian than Oppenheimer: intellectual and moral maturity is still to be achieved through the exercise of criticism in its various modes, but the a priori has become historicized. Bethe's world is still premised on Enlightenment ideas of knowledge, reason, truth, progress, and he harbors strong hopes of universality for them.

Oppenheimer was more "modern" than Bethe, if we interpret being "modern" as referring to sensibility and style. Throughout his life, Oppenheimer was always sensitive to and conscious of style. Modern literature, as a sign of its modernity, at times makes itself exacting. Similarly, Oppenheimer, at times almost willfully, made himself difficult. He also resonated with modernism's "sympathy for the abyss." Irving Howe noted that

> [modernist culture] strips man of his system of beliefs and his ideal claims, and then proposes the one uniquely modern style of salvation: a salvation by, of, and for the self. In the modernist culture, the object perceived seems always on the verge of being swallowed up by the perceiving agent, and the act of perception in danger of being exalted to the substance of reality. *I see, therefore I am.*[20]

The seeing extends to the mind's eye. *I see, therefore I am* is applicable to Oppenheimer, with an emphasis on both *I*, that is, the self, and on *see*, that is, on comprehending and grasping. For much of his life, *I am* did follow from the *I see* thus understood. Oppenheimer also fits being characterized as "modern" if we accept that what distinguishes modern sensibility from earlier sensibilities is that the modern thrives on moral problems and the former on metaphysical problems. Moral problems were always at the center of Oppenheimer's concerns—a legacy of his Jewish and Ethical Culture background.

Isaiah Berlin, in a perceptive essay on "Benjamin Disraeli, Karl Marx and the Search for Identity," commented that those who are born in the solid security of a settled society—as Bethe did—tend to have a stronger sense of social reality: "to see public life in reasonably just perspective, without the need to escape into political fantasy or romantic invention." Those who belong to minorities which are to some degree excluded from participation in the central life of their community—as Oppenheimer was by virtue of his Jewishness—"are liable to develop either exaggerated resentment of, or contempt for, the dominant majority, or else over-intense admiration or indeed worship for it, or at times, a combination of the two, which leads both to unusual insights and—born of overwrought sensibilities—a neurotic distortion of the facts."[21] Trying out different personas in an effort to see which fits best, assuming different roles to achieve centrality contributed to Oppenheimer's restlessness and prevented him from achieving that noble calm that marks Bethe's disposition. He never found his proper place. What Isaiah Berlin said of late nineteenth century emancipated Jews in general applies to Oppenheimer in particular. According to Berlin, these Jews had lost the buttresses of the discipline of their faith and were facing a marvelous, dangerous, and unfriendly new world. Their "over-anxiety to enter into a heritage not obviously [their] own . . . [led] to [an] over-eager desire for immediate acceptance, [and to] hopes held out, then betrayed: to unrequited love, frustration, resentment, bitterness, although it also sharpens the perceptions, and, like the grit which rubs against an oyster, causes suffering from which pearls of genius sometime spring."[22]

Until his downfall, Oppenheimer was also "modern" in the Baudelerian sense. As a young man, in one of his many facets, Oppenheimer had groomed himself to grasp the "heroic" aspect of the present moment and had cultivated the will to "heroize" the present. After the war, I would suggest, there were also attempts by Oppenheimer to estheticize politics, so that at times political activities became "an ascetic elaboration of the self" worked out by a kind of Nietzchean *übermensch* in pursuit of his own desires.[23]

Oppenheimer was deeply conscious of the "discontinuity of time" in physics, in literature, in the arts, and in politics. In his lectures after the war, he often spoke of the break with tradition, of the feeling of novelty he was experiencing in the new world he was living in and of the vertigo brought about by the tempo of change.[24] I would characterize him as almost "postmodern" during the last decade of his life. He then saw modernism as a localized, historically contingent state of consciousness,[25] and became "very suspicious of statements that refer to totality and completeness" and to the eternal and the immutable: "Only a malignant end can follow the systematic belief that all communities are one community; that all truth is one truth; that all experience is compatible with all other; that total knowledge is possible; that all that is potential can exist as actual.[26] He likewise became very distrustful of order, "which is hierarchical in the sense that it says that some things are more important than others—that some things are so important that you can derive everything else from them."[27] As far as science was concerned, "No part of science follows, really from any other in any usable form. I suppose nothing in chemistry or in biology is in any kind of contradiction with the laws of physics, but they are not branches of physics. One is dealing with a wholly different order of nature."[28] He believed that no theory in science is ever closed or finished: "Science is always limited, and is in a profound way unmetaphysical, in that it necessarily bases itself upon the broad ground of common human experience, tries to refine it within narrow areas where progress seems possible and exploration fruitful. Science is novelty and change. When it closes it dies.[29] And he was contemptuous of the claim of the logical positivists that only within the context of the natural sciences can one speak of truth: "They limit very much the meaning and the scope of what it is worth talking about; and they pre-empt the word 'truth' rather harshly for the content of the sciences. That need not bother us. One does not have to insist that the poet speaks of truth; he does sometimes; most of the time he is doing something equally, perhaps more important. He speaks meanings, and he speaks order."[30]

While nurturing his various talents and proclivities, but conscious of his fracturedness, Oppenheimer yearned for wholeness and a more integrated self. In 1930 he wrote his brother, Frank, of his longing for maturity: "In mature people there comes to be more and more of a certain unity, which makes it possible to recognize a man in his most diverse operations, a kind of specific personal stamp, which characterizes not so much the what as the hows of a man's business.[31] One could portray the *young* Oppenheimer in terms of Baudelaire's depiction of the "modern" painter Constantin Guys:

Be very sure that this man, such as I have depicted him—this solitary, gifted with an active imagination, ceaselessly journeying across the great human desert—has an aim loftier than that of the mere *flaneur*, an aim more general, something other than the fugitive pleasure of circumstance. He is looking for that quality which you must allow me to call "modernity"; for I know of no better word to express the idea I have in mind. He makes it his business to extract from fashion whatever element it may contain of poetry within history, to distill the eternal from the transitory.[32]

How to find the elements of poetry within history, how not to be reduced to mere history, how not to be a rebel yet fix a limit to history[33]—these are the questions that delineate some of Oppenheimer's struggles with himself. And perhaps Nietzsche's insight provides a partial answer as to why Oppenheimer never resolved his internal conflicts:

One who cannot leave himself behind on the threshold of the moment and forget the past, who cannot stand on a single point, like a goddess of victory, without fear or giddiness, will never know what happiness is; and worse still, will never do anything to make others happy. The extreme case would be the man without any power to forget who is condemned to see "becoming" everywhere. Such a man no longer believes in himself or his own existence; he sees everything fly past in an eternal succession and loses himself in the stream of becoming. At last, like the logical disciple of Heraclitus, he will hardly dare to raise his finger.[34]

One might characterize Oppenheimer until 1936—when he met Jean Tatlock[35] and also became politically engaged—as a "dandy" in the Baudelairean usage of that word.[36] For Baudelaire, "dandy" implied "a quintessence of character and a subtle understanding of the entire moral mechanism of the world; with another part of his nature, however, the dandy aspires to insensitivity." The desire for insensitivity for Oppenheimer was complex. On the one hand, it stemmed from his hypersensitivity; and on the other, from a need to assert his superiority. In 1929, Oppenheimer had written to his brother, Frank, that "it is not easy for me—to be quite free of the desire to browbeat somebody or something."[37] He still was not free of this trait after the war, though, interestingly, it was not in evidence during the Los Alamos years.[38] Abraham Pais, who was his colleague at the Institute for Advanced Study, commented that Oppenheimer would anger him at times "by his arrogance if not cruelty when a young academic would not clarify or miss a point, cutting him down with unnecessary biting comments."[39]

How to control all his conflicting feelings as well as his prodigious talents had led him already as a young man to confront various forms of obedience. In 1932, in a letter to his brother, Frank, he described the virtue of discipline:

I believe that through discipline, though not through discipline alone, we can achieve serenity, and a certain small but precious measure of freedom from the accidents of incarnation, and charity, and that detachment which preserves the world which it renounces. . . . But because I believe that the reward of discipline is greater than its immediate objective, I would not have you think that discipline without objective is possible: in its nature discipline involves the subjection of the soul to some perhaps minor end; and that end must be real, if the discipline is not to become factitious. Therefore I think that all things which evoke discipline: study, and our duties to men and to the commonwealth, war, and personal hardship, and even the need for subsistence, ought to be greeted by us with profound gratitude; for only through them can we attain to the least detachment; and only so can we know peace.[40]

The subject of discipline and of the conduct and obligations of virtuous men is a recurring theme in Oppenheimer's public addresses after World War II. In a revealing and clearly very meaningful address delivered in 1957 to the undergraduates at Cal Tech, Oppenheimer came back to the relevance of discipline in the new world, where young people have to find their way "into an immense cognitive jungle . . . with very little guide, either from synoptic kinds of knowledge . . . which say: This is important; this is unimportant . . . or from the state of the world, which doesn't, in any clear or loud voice, say: Learn this; ignore that." The world of knowledge had changed. The metaphor that had characterized the world he, Oppenheimer, had inherited as a youth was that of a finite, exhaustible chamber. The new world, however, is essentially infinite, knowable in many different ways; and since it is infinite, only partial knowledge, always supplementable and never closing, is possible. But "all these paths of knowledge are interconnectable; and some are interconnected, like a great network—a great network between people, between ideas, between systems of knowledge—a reticulated kind of structure which is human culture and human society."[41] And to underscore the fact that he did not contemplate with abhorrence this new condition which some had described as chaotic, and that he did not despair "because of the absence of global traits to our knowledge," Oppenheimer read a poem to the students that seemed to him to fit a little "not only with this general situation, but perhaps even with the local situation." The poem was George Herbert's *The Collar*, a poem "he liked" and one that meant a great deal to him:

> I struck the board and cry'd, No more.
> I will abroad.
> What? shall I ever sigh and pine?
> My lines and life are free; free as the road,
> Loose as the winde, as large as store.
> Shall I be still in suit?
> Have I no harvest but a thorn
> To let me bloud, and not restore
> What I have lost with cordiall fruit?
> Sure there was wine
> Before my sighs did drie it: there was corn
> Before my tears did drown it.
> Is the yeare onely lost to me?
> Have I no bayes to crown it?
> No flowers, no garlands gay? all blasted?
> All wasted?

Not so, my heart: but there is fruit,
                And thou has hands.
        Recover all thy sigh-blown age
On double pleasures: leave thy cold dispute
Of what is fit, and not. Forsake thy cage,
                Thy ropes of sands,
Which pettie thoughts have made, and made to thee
        Good cable, to enforce and draw,
                And be thy law,
While thou didst see wink and wouldst not see.
                Away; take heed:
                I will abroad.
Call in thy deaths head there: tie up thy fears.
        He that forbears
        To suit and serve his need,
                Deserves his load.
But as I rav'd and grew more fierce and wilde
        At every word,
Me thoughts I heard one calling, *Child!*
        And I reply'd, *My Lord.*[42]

The poem, a cry of despair and for freedom, must have baffled the students, for it is difficult poetry, written in a somewhat archaic language.[43] Oppenheimer suffered from bouts of deep depression, and Herbert's poetry resonated with this deeply private person, who was not given, and was perhaps unable, to show his feelings, but who was hungry for simple comradeship.[44] Elizabeth Clarke, in her study of Herbert's poetry, noted that in the technical language of the day, many of Herbert's poems, and *The Collar* in particular, are "ejaculations"—at times shrieks of ecstasy, but more frequently cries of affliction, sorrow, and despair.[45] In the poems, these were followed by "motions" of the spirit that bring about momentary divine bliss. Herbert was a cleric when he wrote *The Collar*. At the time, "collar" was in common use to express discipline; preachers would use the word for the obedience imposed by conscience. It has been suggested that the poem represents in psychological terms the events of the Christian moral drama—the Fall, the Atonement, and the Redemption.[46] The speaker of the poem rails against discipline. But in reaching for the "fruit," as did Adam, he simultaneously reaches for the fruit of the Cross, and his rebellion is to be finally overcome by the sacrament of the Eucharist. "Part of the brilliance of the poem lies in the fact that it expresses rebellion and atonement in the same vocabulary, and in so doing epitomizes its central idea: that rebellion necessarily entails, because of God's justice and mercy, atonement."[47]

The poem's grip on Oppenheimer might indeed have been that it addresses the issue of sin and expiation. With his act of defiance, the building of the atomic bomb, he had eaten of the tree of knowledge. How to atone for the blood on his hand caused by its use became an obsession with Oppenheimer.[48] His apathetic stand at his trial in 1953 probably in part reflected his need to do penance. In 1957, when he read the poem to the students, its theme of despair, discipline, and rebellion clearly still resonated with his inner concerns, but his sensitivity to its message perhaps admits a somewhat less Christian interpretation—though the fact that Oppenheimer should identify so deeply with a poem that is Christian at its core should be noted. The Christian God who can only reveal himself by becoming human, is a God for whom no mediation is possible. According to Christian theology, the moral disorder entailed by Adam and Eve's rebellion of eating the fruit of the tree of knowledge could only be overcome by Christ's sacrifice and his gift of *caritas*. But for Oppenheimer, the chaos and moral disorder entailed by the new knowledge—that of the atomic and nuclear world, that of the new biological world—could only be overcome by "what companionship and intercourse and an open mind can do. . . . [For t]he greatest relief and opening is comradeship, and that ability to learn from others of what their world is like." Salvation lies in tolerance and in reverence for knowledge and skill. Oppenheimer counseled the students: "If you have learned how to be something, how to be a competent professional, you will know a great deal about what is good in the world. You will have a bond in common with every other man who is a scholar or a scientist." Note that on leaving Los Alamos in 1945 he had characterized as the deepest bond "the common bond with other men everywhere." In his address to the students in 1957 he was content with less, namely with establishing "a bond in common with every other man who is a scholar or scientist."

It was part of Oppenheimer's tragedy that, after World War II, he felt that he no longer was a creative scientist and that he therefore had lost part of his "anchor in honesty," and hence integrity. George Kennan, who got to know Oppenheimer after the war[49] and became his colleague at the Institute in 1951, made some of the most insightful observations of Oppenheimer's personality. Kennan described Oppenheimer

as in some ways very young, in others very old; part scientist, part poet; sometimes proud, sometimes humble; in some ways formidably competent in practical matters, in other ways woefully helpless: . . . a bundle of marvelous contradictions. . . . His mind was one of wholly exceptional power, subtlety, and speed of reaction. . . . The shattering quickness and critical power of his own mind made him . . . impatient of the ponderous, the obvious, and the platidinous, in the discourse of others. But underneath this edgy impatience there lay one of the most sentimental of natures, an enormous thirst for friendship and affection, and a touching belief . . . in what he thought should be the fraternity of advanced scholarship . . . [a belief that] intellectual friendship was the deepest and finest form of friendship among men; and his attitude towards those whose intellectual qualities he most admired . . . was one of deep, humble devotion and solicitude.[50]

The greatest tragedy of Oppenheimer's life was not the ordeal he went through over the issue of his loyalty, but his failure to make the Institute for Advanced Study a true intellectual community. As Kennan noted, Oppenheimer was often discouraged, and in the end deeply disillusioned, by the fact that

the members of the faculty of the Institute were often not able to bring to each other, as a concomitant of the respect they entertained for each other's scholarly attainments, the sort of affection, and almost reverence, which he himself thought these qualities ought naturally to command. His fondest dream had been [Kennnan thought] one of a certain rich and harmonious fellowship of the mind. He had hoped to create this at the Institute for Advanced Study; and it did come into being, to a certain extent, within the individual disciplines. But very little could be created from discipline to discipline; and the fact that this was so—the fact that mathematicians and historians continued to seek their own tables in the cafeteria, and that he himself remained so largely alone in his ability to bridge in a single inner world those wholly disparate workings of the human intellect—this was for him [Kennan was sure] a source of profound bewilderment and disappointment.[51]

Bethe has been more modest. Furthermore, Bethe never stopped doing physics, never ceased "preserving his competence and mastery of his profession"[52] and therefore never lost his anchor in integrity. To the present, he works every day on physics. His scope may have narrowed, but he has maintained full mastery of that terrain. And at Cornell he created an intellectual and social community within the

Laboratory of Nuclear Studies that gave him a warm and harmonious fellowship—of the mind and of the heart.

Baudelaire captured something of Bethe's genius. Baudelaire interpreted genius "as nothing more than nor less than childhood recovered at will" and characterized a genius as "a person for whom no aspect of life has become *stale*. . . . [He is] a master of that only too difficult art—sensitive spirits will understand me—of being sincere without being absurd."[53]

2.

J. ROBERT OPPENHEIMER

> The greatest difference between the poet and the
> ordinary person is found . . . in the range, delicacy,
> and freedom of the connections he is able to make
> between the different elements of his experience.
> —I. A. Richards (1952)

OPPENHEIMER AND THE ETHICAL CULTURE MOVEMENT

Robert Oppenheimer was born in New York on April 22, 1904, into a
well-to-do, emancipated Jewish family of German descent. His father,
Julius, was a successful businessman who had emigrated to the
United States in 1888 when he was seventeen years old to work in
the textile-importing business of relatives. His mother, Ella Friedman,
was an artist whose family had come from Germany to Baltimore in
the 1840s. A brother, Frank, was born in 1912. The Oppenheimers
lived on Riverside Drive near 88th Street in a large eleventh-floor
apartment overlooking the Hudson River. Although Jewish, the Oppenheimers had no temple affiliation. During Robert's youth, they
were active members of the Ethical Culture Society that Felix Adler
had founded in 1876, and Julius was a member of the Board of Trustees of the Society from 1907 to 1915. It was therefore natural for the
young Robert to be sent to the Ethical Culture School, which was
housed in a building adjacent to the Society's headquarters on Central Park West near 63rd Street. He entered its second grade in September 1911 and graduated its high school in February 1921. Ethical
Culture thus played a significant role in the young Robert's life, and it
is therefore important to understand its role in shaping his character.[1]

At the center of the Ethical Culture movement from its founding in
1876 until his death in 1933 stood Felix Adler. He had come to the
United States in 1857 at age six when his father, Samuel Adler, a rabbi
in a small town in Rhenish Hesse in southwest Germany, accepted the
pulpit of the largest Reform congregation in the United States, Temple
Emanu-El in New York, most of whose members were wealthy Jews of
German descent.[2] Together with Isaac M. Wise and Max Lilienthal, the
learned Samuel Adler became one of the leaders of American Reform

Judaism. Felix had been expected to succeed in his father's position, but during his studies in Germany from 1870 to 1873 his horizons widened by attending the lectures of Hermann Cohen,[3] Hermann Bonitz, and Eugen Dühring in Berlin. He also became exposed to the textual and historical criticism of the Old and New Testaments by Abraham Geiger, Ernest Renan, and David Friedrich Strauss and studied Darwin's theory of evolution and its implication for religion. He thereafter could no longer accept the basic creeds of the Reform movement: belief in a personal God and in the special mission of the Jews. The latter dogma, one of the central tenets of Reform Judaism, asserted that Jews had been chosen by God to teach mankind monotheism and to inspire the human race to be morally prepared for the coming of messianic times of universal peace and universal fellowship.[4]

In Germany, Adler committed himself to becoming "the minister of a new religious evangel." He vowed that "instead of preaching the individual God [he would] stir men up to enact [Kant's] Moral Law . . . [and would] go out to help arouse the conscience of the wealthy, the advantaged, the educated classes, to a sense of their guilt in violating the human personality of the laborer."[5] Upon his return from Germany, in a sermon delivered at Temple Emanu-El in October 1873, he outlined his vision of the "Judaism of the Future." The world is changing rapidly and dramatically, Adler declared; if Judaism is to survive it must shed its "narrow spirit of exclusion" and become a universal humanitarian religion with deeds and not creed as its foundation. The deeds were to be channeled into practical social activities, and in particular, the moral and intellectual education of the laboring masses.[6] Three years later, breaking his ties with the Jewish communion, Adler called this religion "Ethical Culture." The precipitating event was the nonrenewal, in 1876, of his appointment as a nonresident professor of Semitic and Oriental literature at Cornell University. Adler, with the support of some of his friends and patrons from the German Jewish community, then initiated a Sunday lecture movement, which subsequently developed into the New York Society of Ethical Culture. The meetings took place on Sunday mornings and attracted a large audience made up principally of members of Temple Emanu-El and of unaffiliated, essentially assimilated, Jews. Though some organ music was played at these Sunday gatherings, "to elevate the heart," no prayers were recited nor was any religious ceremony performed; the major part of the assembly was taken up with a lecture by Adler.

Whereas for Adler the founding of the Society for Ethical Culture, and his concomitant rupture with the Jewish fellowship, was the result of an intellectual confrontation with the tenets of Reform Judaism, for many of the approximately four hundred Reform Jews who joined the movement, who for the most part had been born in Germany, the appeal of the Society was more pragmatic: it offered a solution to the problems presented by the new context in which Jews found themselves in the United States after the Civil War.

The Jews who had emigrated to the United States during the first half of the nineteenth century had come predominately from Central Europe, many of them from German-speaking states in the aftermath of the failure of the 1848 revolutions. On the whole, they had prospered, become well-to-do, some very wealthy.[7] By the early 1870s, those among them who identified themselves as Reform Jews believed they were ushering in a new era of universal brotherhood in which, in the words of Rabbi Isaac Mayer Wise, "the whole human race shall be led to worship one Almighty God of righteousness and truth, goodness and love."[8] However, by the late 1870s this optimism had faded.

Although there had been expressions of "ordinary" anti-Semitism—that is, bigotry and hostility in daily life, and discrimination in residence, employment, education, and social clubs—since the founding of the Republic that had given them their emancipation,[9] American Jews had believed that prejudice and intolerance would disappear as the nation grew more prosperous and its democratic institutions matured. But these hopes were dispelled in the aftermath of the Civil War and the crash of 1873, when discrimination of Jews became more extensive and more organized.[10] Fierce anti-Semitism had also broken out in Germany in 1873 in the wake of the stock market crash there that had punctured the speculative era following German unification. It culminated with a petition being presented to Bismarck in November 1880, asking that the immigration of foreign Jews into Germany be halted; that Jews be barred from holding positions of authority; that Jewish teachers not be permitted to instruct in public schools; and that a vigilant surveillance of the Jewish population by means of official statistics be reinstated.[11] That racially based anti-Semitism was not only condoned, but in fact fostered in the highest intellectual and most influential German political circles,[12] came as a great shock to the German Jews of the Reform movement.[13] Germany had represented emancipation and enlightenment. Felix Adler in the first of two lectures devoted to these events declared that they constituted an "experience for which we were not prepared, and which overwhelms the spirit with sadness."[14]

Two well-publicized incidents were indicative of the change that was taking place in the United States.[15] In 1877 Joseph Seligman, the wealthiest Jew of German descent in New York, an international banker who wielded great influence in Washington, a staunch supporter of Felix Adler's activities and one of his most generous benefactors,[16] was refused accommodations at the Grand Union in Sarasota Springs, a luxury resort under the control of Judge Henry Hilton, a hotel in which he and his family had often stayed during past summers. And in 1879, Austin Corbin, the president of the Long Island Railroad and of the Manhattan Beach Company, which was trying to develop Coney Island into a fashionable summer resort, announced that "we do not like Jews as a class" and that they would not be welcome at his hotel, since he was convinced "that we should be better off without than with their custom."[17] These incidents had a profound psychological effect on Reform Jews, for they indicated the limits to their upward mobility.

The Seligman-Hilton incident proved to be the first among a swell of anti-semitic incidents. Anti-semitic statements began appearing in the press, and social discrimination of Jews became more prevalent. The German Jewish population's apprehension was intensified when, to silence the threats of Darwinism and biblical criticism, evangelical Protestant denominations called for a "Christian America." Even as liberal and progressive an individual as Andrew White characterized nondenominational Cornell as a "Christian" university. The unease became further aggravated in 1881 upon the assassination of Tsar Alexander II by revolutionary terrorists, which led to a wave of pogroms throughout Russia, largely instigated by agents of the government. The virulent forms of anti-semitism that became sanctioned by the new tsarist regime in turn led to the first wave of massive immigration to the United States of "Ostjuden."[18] The presence of these apparently unassimilable Jews who seemingly threatened to outnumber the older Anglo-Saxon, northern stock, nurtured the racial anti-Semitism that had been promulgated by Gobineau. Petitions for restricting the immigration of Jews were circulated, and a bill with that end in mind was passed by Congress but vetoed by Grover Cleveland.[19] Social discrimination at all levels became widespread, and by the end of the century restrictive agreements barring Jews from certain residential areas were common. Jews found it difficult to have access to Gentile private schools, to become faculty members in the leading American universities, and, most significantly, to be employed by most of the new large industrial corporations.

The post-Civil War period had witnessed the formation and growth of the institutional framework that defined the official culture of society, that is, of the new wealthy class: "museums, symphony orchestras, opera houses, learned societies; and an array of journals and publications that passed judgment on taste and ideas. This apparatus was in the hands of a few key figures, the critics and influential scholars who were the ultimate arbiters"; and at its core was the university.[20] Jews found themselves being barred from these institutions, and their exclusion from the restricted WASP prep schools denied them access to the elite universities and through those access to the upper echelons of society.

In the face of these deeply disturbing developments two paths were open for liberal Reform Jews to take from the mid-1870s on. One was to assimilate; the other was to undertake a revitalization of the Reform movement and to give new meaning to being a Reform Jew. Both were taken.[21] The Ethical Culture movement became a way station to complete assimilation for many of the moneyed, European-born, first-generation American Reform Jews who joined it.[22] They could assert that they were Americans whose religion was Ethical Culture. But for the most part, the next generation, being American born and more acculturated to their surroundings, did not look upon Ethical Culture as a "religion." One reason for this was that the outside world often still identified them as Jews. Another was that their secularism took on a much greater importance: they did not need any religion.

THE AGENDA OF THE ETHICAL CULTURE SOCIETY

Felix Adler was a member of the generation that drew its inspiration from philosophers who tried to find, as John Dewey was to say of William James, "a *via media* between natural science and the ideal interests of morals and religion" and who in the process formulated an ethics of rational benevolence.[23] But these thinkers had failed to articulate a political philosophy suited to the dramatic transformation of the industrial landscape of the end of the nineteenth century. Adler's generation extended and broadened their mentors' ideas of democracy by developing their predecessors' analysis of knowledge, duty, and responsibility. It is that generation that transformed socialism into social democracy and the ideology of classical liberalism into the tenets of progressivism and helped devise the welfare state. The connection between these two generations is evident in Adler's writings and comes sharply into focus in the works of John Dewey, whose intellectual trajectory and educational interests paralleled those of Felix Adler. But whereas James and Dewey made "the transition between the vocation of the philosopher as preacher and the profession of philosopher as academic,"[24] Adler remained a transitory figure in the professional arena—he remained a part-time preacher all his life and became only a part-time academic. He has become a transitory personage in the history of education, of philosophy and of religion, even though his educational innovations—such as the Workingman's school—anteceded those of Dewey, and much of his social creed was taken over by the social gospel movement.

Already in his 1873 Temple Emanu-El sermon, Adler had expressed his concern about the status of laborers and of women and children in modern industrial societies. Adler's emphasis on the condition of the working classes stemmed from his reading Friedrich Albert Lange's *Die Arbeiterfrage*.[25] He later recalled that this book, "while not great, . . . opened for me a wide and tragic prospect, an outlook of which I was until then in great measure oblivious, an outlook on all the moral as well as economic issues involved in what is called the Labor Question." Although Adler had not accepted Hermann Cohen's claim that "if there is to be anything like religion in the world hereafter, Socialism must be the expression of it" he did believe that there was "truth in what [Cohen had] said, and [that] I must square myself with the issues that socialism raises. Lange helped me do that."[26] He read Karl Marx's *Kapital* soon after it was published, and he was "profoundly stirred by the chapters . . . in which [Marx] collect[ed] from the English Blue Books frightful evidence of the mistreatment of laborers and especially of children in the early part of the nineteenth century." He was sympathetic with the objectives of Marxian socialism, but took exception with its view that moral ideas were epiphenomena or by-products of economic development.[27]

The improvement of the lot of laborers, that of women and children, were among Adler's primary concerns throughout his life; and the amelioration of the conditions of the laboring classes became one of the cardinal commitments of the Ethical Culture movement he spearheaded. Many of the lectures Adler delivered at the Sunday meetings addressed these issues. In the first set of lectures he gave in the fall and winter of 1876 he presented a critique of the traditional tenets of Judaism and Christianity. A subsequent set traced the history of human aspirations and advanced "a standard of duty for today." One of the lectures in that series dealt with "The Enlightenment of the Masses." In it Adler asserted that this was "the problem on which the future of society depend[ed]" and declared that liberal religious institutions must become potent forces for enlightenment not just on Sundays but on the other six days as well. "Liberalism must pass the stage of individualism, [and] must become the soul of great combinations"; moreover, it must organize to produce "institutions grounded in the needs of the present." In a later lecture he declared that the tenements in which workers and their families lived and the appalling conditions under which they toiled demonstrated that justice was being trampled by indulging in Lincoln's dream wherein "the laborer of today expects to become the employer of tomorrow." The challenge now was "to raise them in their capacity as laborers," and this meant "just remuneration, constant employment, and social dignity."[28]

A year later, in 1877, keeping with his slogan that the Ethical Culture Society commit itself to "Diversity in the creed, unanimity in the deed," Adler initiated two projects to address "the labor problem." The first sent visiting nurses to the homebound sick in poor neighborhoods; the second started a free kindergarten for the children of workingmen, the first of its kind in the East. Adler's focus on the education of working-class children reflected the widespread efforts to clarify the form and functions of schools in the wake of the Civil War and of the dramatic growth of the urban population due to industrialization, public health measures, and immigration. A sense of communal disintegration, of drift, and of lack of standards was widespread. The shock of the Civil War, the impact of industrialization and urbanization, the creation of gargantuan fortunes in the hands of a few, the erosion of traditional religion in the face of higher criticism and Darwinism, all were factors in the collapse of the communal institutions that had bound the nation together. Under these circumstances, people turned to the schools to supply the guides to action that the individual could no longer acquire through the slow accretion of experiences. They were to impart the instruction that the family and the church had offered in more stable societies and that were now simply lacking.[29]

The efforts of Adler and of Dewey to reform primary and secondary education had their roots in the new context the nation found itself in after the Civil War. Adler's kindergarten project proved so successful that in 1880 it was expanded into an eight-grade, tuition-free elementary school, called the Workingman's School. Even though its name was the Workingman's School, admission to it was not limited to children from workers' families. Its aim was to be a model school.[30] It was to furnish an elementary education to pupils from all social classes which would prepare them for life and equip them to earn a living in a rapidly developing industrial society. Adler personally supervised the hiring of the teachers, making sure that only the most competent would be employed.[31] Besides instruction in the usual subjects of arithmetic, reading, writing, history, and geography, the school also provided a diversified program of arts and crafts, indoor and outdoor activities, drama, dance, and gymnastics.[32] Each grade also received instruction in morality.

One of the striking features of the curriculum of the Workingman's School was the inclusion of manual training, hands-on "industrial education" in the primary grades. But whatever vocational aspect there was to the enterprise soon gave way to the cultural implications of manual training.[33] Adler considered manual training as an effective means to demonstrate the dignity of labor and to inculcate a love of labor. Manual labor would also indicate the organic solidarity of the laboring and managerial classes and thus bridge social divisions.[34] Equally important, Adler saw manual training as an effective way of disciplining the will and a way of teaching the interrelation between thought and action. Every act involves the coordination of will, reason, and feeling. The respective weight given to these factors differentiated different educational philosophies.[35] Adler always championed reasoned control rather than romantic assertiveness or self-expression: "The will may be compared to the power that propels a ship through the waves. Feeling is the rudder. The intellect is the helmsman."[26] Moreover, reasoned control could often be encapsulated in moral rules.[37]

In his address at the opening of the Workingman's School, Adler declared:

> We propose to give [the children] that which will secure them
> bread thereafter, and many of the higher treasures of human ex-
> istence, we hope besides; we propose to give them a broad and
> generous education, such as the children of the richest might
> be glad in some respect to share with them which will prepare
> them for their future station in life, but also make them capa-
> ble of living in a truly human way.[38]

His statement proved prophetic. By 1890 the school was so successful and had attracted so many students that the cost of running it could no longer be met by the contributions from the Society. Affluent Jewish members of the Society who where finding the doors to WASP private schools closed, and who wanted their children to have the benefits of the school's outstanding education, then proposed having their children admitted to the school as tuition-paying students. This new policy was instituted, and it allowed the school to add a high school in 1895. At that same time, the name of the school was changed to The Ethical Culture School and moved to Central Park West in a building not far from the Society's headquarters and meeting place.[39] The admission of tuition-paying students in many ways transformed the school. Although the school kept on supporting about 10 percent of its student body with scholarships, it became a model preparatory school for very bright students from rich families who would go on to the best colleges in the country. Mabel Neumann, a historian of education who attended the school, noted that the move to Central Park West signaled alteration of Adler's tactics in effecting reconstruction of the poor. The Workingman's School was designed to build the strength of each poor boy and girl educated in it. The Ethical Culture School, whose students were children of members of the Ethical Culture Society, and, perforce, were not poor, was to be infused with the ideal of Ethical Culture. In a word, they were to be reformers, or, as Adler came to call them, "leaders" [persons who will be competent to change their environment to greater conformity with moral ideals]. . . . It was as though Adler had returned to the years 1875–1880, when he was a "minister" of the social gospel.[40]

THE TEACHING OF ETHICS AT THE SCHOOL

In the fall of 1931, two years before his death, Adler addressed the students of the Ethical Culture School.[41] In his talk he summarized its aims: "What the school is for is to make a better world. . . . In order to do that, to make yourself fit for such an undertaking, it is necessary that you should make something of yourself. You must cultivate your best talents in relation to those of others."[42] But this personal development should take place with a concomitant awareness of the interconnectedness of all human beings. Therefore the school was also to be an "educational temple . . . to train for the growing life of the world, . . . an altar to the . . . unknown, unpredictable, inconceivable divine things that slumber as yet unborn in the bosom of mankind."[43] This dual conception had been present ever since Adler founded the school and guided its subsequent growth. The emphasis on making "a better world," rather than a better nation, is what differentiated the Ethical Culture School from other progressive schools. Both sets of schools tried to indoctrinate their students with "political and moral axioms and principles" that would guide their actions as citizens; but at the Ethical Culture School, education for citizenship never became identified with patriotism. The means by which the democratic spirit was inculcated was by making sure that the school would not become a "class" school. "The children of the rich and of the poor, and those of different races were to meet there and learn to respect one another, both in their work and play."[44]

In order to teach students to live a moral life and to strive toward ethical perfection, and do so without grounding ethics on denominational religion, Adler had turned to Kant. Adler accepted the Kantian view that education must be concerned with the intellectual mastery of nature, the glorification of life in art and with its consecration in morality. In the moral sphere, Adler's point of departure was Kant's "categorical imperative"—to treat every person as an end and not as a means. The evolution of the way ethics was taught at the school mirrored Adler's interpretation, and subsequent reformulation, of the Kantian imperative and of the ethics of "practical reason." The method used at the Workingman's School was presented in Adler's book on the "unsectarian" moral education of children. He there indicated that ethics could be taught in a positivistic way:[45]

> Ethics is a science of relations. The things related are human interests, human ends. The ideal which ethics proposes to itself is the unity of ends, just as the ideal of science is the unity of causes. The ends of the natural man are the subject-matter with which ethics deals. . . . The ends of the natural man are to be respected . . . so long as they remain within their proper limits. The moral laws are formulas expressing relations of equality or subordination, or superordination.[46]

As a consequence of reading Edward Tyler, John Lubbock, and other evolutionary anthropologists, Adler had accepted Johann Herbart's "culture epoch" theory,[47] which stipulated that children reproduce in their own development the main stages of evolution which have been passed through by the human race. Herbart had then drawn from the theory the pedagogical inference that the products of the various stages of human evolution are the most appropriate material for children in their corresponding stages of development. Herbartian educational philosophy thus made literature the basis of the curriculum, because it is in literature that the products of the development of the species at its successive stages have been principally conveyed to us.[48] Adler adopted Herbart's doctrine to teach moral principles. He considered each stage to have its specific set of duties and responsibilities, and argued that in each period there is one duty of paramount standing around which the others can be grouped. Furthermore, Adler assumed that the new set of duties at each phase encompassed the previous ones. In order to implant in the students these sets of duties—duties to self and others, which followed Kant's *Tugendlehre* rather closely—Adler selected the reading materials for the successive stages in the student's development from the literature of the corresponding epochs in the life of humankind, such as Homer's *Iliad* and *Odyssey*, the Old Testament, and the New Testament.[49]

Toward the end of the 1890s, Adler became critical of the categorical imperative and compared it to the commands of the corporal in the Prussian army of Kant's times: "Kant's Categorigal Imperative comes to us with the impact of a blow to the head. 'Thou shalt.' Why? We are forbidden to ask that question." Adler came to regard Kant's ethics as "a species of physics," a physics with an atomic, individualistic metaphysics, and to believe that one could not base the science of ethics on an analogy with the science of physics. For the

> manifold with which science deals, which it is its business to unify, is given in sensation, in experience. The manifold with which ethics deals is not given, not supplied at all from without, but is a purely ideal manifold. . . . The organic ideal is that of an infinite system of correlated parts, each of which is necessary to express the meaning of the whole, and in which the whole is present as an abiding and controlling force. The ethical ideal is produced by applying this purely spiritual conception of an infinite organism to human society. To act as if my fellow-beings and as if I myself were members of such a system in which the manifold and the one are wholly reconciled is to act morally.[50]

He then reformulated the categorical imperative as follows: "So act, not as if the rule of thy action were to become a universal law for all rational beings . . ., but so act that through thine action the ideal of an infinite spiritual organism may become more and more potent and real in thine own life and in that of all thy fellow beings"[51] This transcendental aspect is what differentiated Adler's ethics from Dewey's. Also, democratic ideals fit into Adler's philosophy differently than into Dewey's. For Dewey it was the point of departure, whereas Adler never shed a *noblesse oblige* attitude that became a component of the ethos of the Ethical Culture School.[52] Robert Oppenheimer became inculcated with doing the "noble thing" there, a stance that surely resonated with the values that permeated his parents' home. Thus, in a lecture "On Science and Culture" delivered in 1962, Oppenheimer noted that "in those high undertakings when man derives strength from and insight from public excellence, we have been impoverished. We hunger for nobility, the rare words and acts that harmonize simplicity with truth."[53]

Adler's views were incorporated into the teaching of ethics at the school. Since he believed that students should be active recipients of learning and was sensitive to the needs of adolescents, ethics at the high-school level were taught in a seminar that met once a week. No marks were given, and discussions and not lectures were the norm. The problems considered were real, concerned with specific social and political issues the students faced or were about to confront. Adler called these classes "Education in Life Problems." Friess lists some of the topics addressed in the seminars during the first decade of the century. During their first year in the high school, students discussed the ethics of wealth and of poverty, the Negro problem, and the unalienable worth of every human being. In addition, the school aspired to awaken in the students "the spirit of social services by enlisting [their] interest in the work of the settlements and neighborhood houses with which the school [was] in touch."[54] The theme for the sophomores was "The school and family as social organisms," and in particular, the ethics of authority and of gender relations. In the seminar for the seniors, the principal subject was "the State [as] having for its ideal aim the unification of the various vocational groups and the expression of the national character and genius." Other topics listed for the discussions of the senior class were "political ethics: e.g., ethics of loyalty, . . . of relation of church to state." During World I and after the armistice was signed, the period when Oppenheimer was a student at the school, the discussions included the ethics of war and the ethics of peace.[55]

THE MATURATION OF OPPENHEIMER

I have presented the ethos of the Ethical Culture Society in some detail because it molded Oppenheimer's early intellectual development and left, I believe, a deep imprint on him, particularly on his moral outlook. The education Oppenheimer received at the Ethical Culture School shared many of the ideals that had been incorporated into the notion of *Bildung*. To become a *Bildungsträger* required the development of the whole person through the nurturing of the growth of the intellectual faculties, esthetic sensibilities, and moral cognizance.[56] Bethe was molded in that tradition in the *Gymnasium* he attended.

In his interview with Thomas Kuhn on the history of quantum physics, Oppenheimer recalled that when he was ten or twelve, "minerals, writing poetry and reading, and building with blocks—architecture—were the three themes that I did." He disavowed that they had any relation to school and claimed to have done them "just for the hell of it." And yet, crystals, rocks, and building blocks were precisely the kind of materials for the hands-on experience that the school fostered in the primary grades. During his high-school days, the school was headed by a remarkable man, David Saville Muzzey, a man of many parts from whom Oppenheimer took a history course in his senior year, and who was a model of the polymath that he himself would aspire to become.[57] Two other teachers among the highly competent faculty greatly influenced him: Augustus Klock in science and Herbert Smith in literature. The school was an ideal place for the precocious, dazzling, but insecure young Robert. He was able to nurture his difference there yet not feel like an outsider. One of his schoolmates remembered him at age fifteen as "still a little boy; . . . very frail, . . . very shy, and very brilliant."[58] He outgrew his shyness, but the insecurity and a deep unhappiness remained. Paul Horgan, with whom he became good friends during a trip to New Mexico in the summer of 1922 before entering Harvard, told Alice Kimball Smith that "Robert had bouts of melancholy, deep, deep depressions as a youngster."

Oppenheimer had to defer entrance to Harvard for a year because he had contracted dysentery during a visit to Europe in the summer of 1921. To help effect a complete recovery, Robert's parents had asked Herbert Smith to go with him on a trip to the Southwest. The trip brought out one of the things that was troubling Oppenheimer, namely, being Jewish. Robert asked Smith to travel with him as his younger brother—a suggestion that Smith firmly turned down. There are indications that this sensitivity to being identified as a Jew, and his feeling himself to be an outsider, remained with him until the mid-1930s.[59] And even after the war, when anti-Semitism had essentially disappeared in the academy, Oppenheimer often used Christian metaphors when addressing moral issues.

At Harvard, a good friend described the diffident young Robert as somewhat precious and arrogant and noted that at the time "he was completely blind to music."[60] But even though he worked like a demon, he found time to cultivate his amity with Paul Horgan and Francis Fergusson, a classmate from the Ethical Culture School who had also gone to Harvard, and to form new lasting friendships—with John Edsall, Jeffries Wyman, and others. There was, however, no time for dating women.

The year following his graduation from Harvard was one of crisis. He went to Cambridge, England, hoping to work with Ernest Rutherford but instead came under the tutelage of J. J. Thomson at the Cavendish Laboratory. "The business in the laboratory was really quite a sham. . . . I was living in a miserable hole," he reminisced to Kuhn. His experiment on the scattering of electrons on thin beryllium targets to measure the electrical conductivity in thin metallic films did not work, and Thomson, by now an old man, was not very helpful. Even though he was terribly excited by the new developments in physics, his failure as an experimentalist raised doubts about the appropriateness of a career in physics. John Edsall remembered Oppenheimer talking to him at length about the papers of Heisenberg and Schrödinger and the meaning of these advances. Edsall at the time was at St.

John's College working toward a Ph.D. in biochemistry. Their friendship, begun at Harvard in 1923, deepened during their year together in Cambridge.

How acutely troubled Robert was at the time is evidenced by an incident that Francis Fergusson later related to Alice Kimball Smith. During a visit to Fergusson in Paris in the late fall of 1926, in the course of one of their customary exchanges about intellectual and personal matters during which Oppenheimer had revealed to Fergusson "his despair over his inept performance in the laboratory and confidences about unsatisfactory sexual ventures," Oppenheimer "suddenly leapt upon Fergusson with the clear intention of strangling him." Since Fergusson was as tall and more solidly built than Oppenheimer, he easily repulsed the attack, but the display of violence made it clear that Oppenheimer was profoundly distressed.[61] Oppenheimer's "tremendous inner turmoil" was also apparent to Edsall upon Oppenheimer's return to Cambridge after the holidays, but in spite of it Oppenheimer "kept on doing a tremendous amount of work" and carried on as if nothing was the matter.[62]

Oppenheimer went on a hiking trip to Corsica and Sardinia with Edsall and Wyman during the spring break of 1927, and everything seemed to be going well. But at the dinner on their last day in Corsica, Oppenheimer suddenly announced that he could not go with them to Sardinia as "he had left a poisoned apple on the desk of Pattrick Blackett at the Cavendish, and must make sure that Blackett was all right."[63] It is possible that Oppenheimer had meant the reason for his having to return to the Cavendish to be taken metaphorically. In any case, when Edsall returned to Cambridge, he found Oppenheimer hard at work on some calculations on X-ray absorption in hydrogen using the new wave mechanics, which Edsall checked for him.[64] Still, Oppenheimer's frame of mind was worrisome enough for someone in authority at Christ's College to write to his parents, who came to Cambridge to visit him. They found him a new psychiatrist in London who they thought would be more helpful than the one he had been seeing in Cambridge. The crisis seems to have abated by itself and was overcome by the end of summer, when Oppenheimer realized that his psychiatrist "was too stupid to follow him and that he knew more about his troubles than the [doctor] did."[65]

The nature of all the factors involved in the crisis in the winter of 1926 have not been disclosed.[66] Fergusson had pointed to Oppenheimer's frustration in his work at the Cavendish, his general unhappiness with the Cambridge culture, and his experiencing "unsatisfactory sexual ventures." Intrigued by the last remark, I asked John Edsall in 1987 whether the crisis could have been a confrontation on Oppenheimer's part with his sexual identity, that questions of sexual polarity had arisen. More specifically, I asked Edsall for his reaction to the suggestion that Oppenheimer harbored marked but latent homosexual feelings—and that this inclination was exacerbated by the Cambridge setting where such feelings were openly displayed.[67] Edsall found the suggestion plausible and consistent with the events during 1926–27, but tempered his acquiescence with the comment "I am certainly no psychiatrist."[68] The issue is of relevance only insofar as it perhaps points to another aspect of this exceedingly complex man, a feature that made cohesiveness and integrity of self more difficult.[69]

Oppenheimer's relations with women appear always to have been conflicted.[70] At Harvard he had very little contact with the opposite sex. It seems likely that his first sexual quests were attempted in England, which he described to his brother Frank as "erotic labours." But it is not immediately apparent how to interpret the advice he gave to the fourteen-year-old Frank in March 1928:

> As for the last rule: Don't worry about girls, and don't make love to girls unless you have to: DON'T DO IT AS A DUTY. Try to find out, by watching yourself, what you really want; if you approve of it, try to get it; if you disapprove of it, try to get over it.—This has all been very dogmatic, and I hope you will forget most of it; but some of it may possibly be of use to you, as the fruit and outcome of my erotic labours.[71]

What seems to be clear is that his meeting Jean Tatlock in the fall of 1936 and their ensuing relationship transformed Oppenheimer. His students and friends noticed the change. His ties with them deepened: he became more compassionate, less cutting and haughty, and more self-confident. It was Jean Tatlock who also made him partake more actively in the political and economic issues of the days: the Loyalist cause in Spain, the plight of migratory workers, that of teaching assistants at Berkeley, and more generally, the devastating consequences of the Depression.[72] In his letter to the Gray Board, which was reviewing his security clearance in 1954, Oppenheimer commented that this novel awareness gave him a "new sense of companionship" and that at the same time he came to feel "that he was coming to be part of the life of my time and country." I would submit that his rewarding relationship with Jean Tatlock allowed him to act on the values that he had been inculcated with at the Ethical Culture School: concern with the labor question, with the conditions of the working class, and with public education. At the Ethical Culture School he had been told that "his personal development must take place with a concomitant awareness of the interconnectedness of all human beings, . . . that liberalism must pass the stage of individualism, [and] become the soul of great combinations, . . . [and that] what the school is for is to make a better world." These tenets were consonant and resonated with all the "left wing" causes he began committing himself to in 1937.

Oppenheimer took up these themes of cooperation and of the interconnectedness of scientists in a radio address on the occasion of Albert Einstein's sixtieth birthday in 1939:

All discoveries in science grow from the work, patient and brilliant, of many workers. They would not be possible without this collaboration; they would not be possible without the constant technological developments that are necessary to new experiments and new scientific experience. One may even doubt whether in the end they can be possible except in a world which encourages scientific work, and treasures the knowledge and power which are its fruits.

Oppenheimer went on to remark that one of the most "spectacular project[s] of contemporary atomic physics was to tap the sources of the sun's energy—a project made possible by Einstein's early work on the theory of relativity," and noted that in this as in every case one would find that the development of Einstein's discovery was "so closely interwoven with the work of countless other scientists and technicians . . . that we would come away [from an analysis of the development] with a deepened conviction of the cooperative and interrelated character of scientific achievement."[73]

With his involvement with Jean Tatlock, his conflict with his Jewishness receded and the issue acquired a new meaning in the face of the vicious anti-Semitism of Hitler's National Socialist Germany.[74] "I had a continuing, smoldering fury about the treatment of Jews in Germany," Oppenheimer told Kuhn. He helped his relatives leave Germany and to emigrate to the United States.[75] Concomitantly, the passion that he could express in teaching and in doing physics found a new outlet. Haakon Chevalier, whom he befriended in the fall of 1937, spoke of "Opje's devotion" to organizing the teachers' union and commented on the fervor he displayed in all his political activity.[76] Chevalier interpreted the importance that Oppenheimer attached to it as reflecting his concern for "the fate of man. It invested his words, and the position he took on political questions, with a peculiar solemnity. This was the 'Hebrew Prophet' side of his nature."[77]

But even though they "were twice close enough to marriage to think of [themselves] as married," Oppenheimer's relationship with

Jean Tatlock did not last. They broke up, and between 1939 and her suicide in 1944 they saw each other but rarely.[78] According to Chevalier, after the breakup Oppenheimer paid "assiduous court to a succession of young ladies, . . . breaking several hearts, and finally late in 1939, broke off with the last of these when he met Katherine Puening Harrison in Pasadena."[79] "Kitty" was the wife of one of his colleagues, Stewart Harrison, an English physician doing cancer research at Cal Tech. She had previously been married to Joe Dallet, a Communist party organizer who had fought with the Lincoln Brigade in Spain and had died fighting there.[80] Oppenheimer married her in November 1940 after her divorce from Harrison. It was evidently a good marriage initially, for Oppenheimer and Kitty were in love and made a fine team. Edsall, who saw Oppenheimer in January 1941 for the first time in over a decade, felt that he was then "a far stronger person, that [the] inner crises that he had been through in those earlier years he had obviously worked out and achieved a great deal of inner resolution of them. I felt a sense of confidence and authority, although still tension and lack of inner ease in some respect."[81] In May 1941 Kitty bore a son, whom the Oppenheimers named Peter. Their daughter, Katherine, was born in December 1944 at Los Alamos.

However, strains developed between Kitty and Robert—the name she insisted everyone call him. The bomb project and the directorship of Los Alamos put enormous demands on Oppenheimer's time and energy. The ensuing stresses were also felt by Kitty, and she developed a drinking problem. Her sister-in-law, Frank's wife, recalled that on one occasion when she visited Los Alamos Kitty invited her for cocktails. "When I arrived, there was Kitty and four or five other women—drinking companions—and we just sat there with very little conversation—drinking. It was awful and I never went again."[82]

The drinking problem evidently persisted after the war, and it was apparent to many after Robert and she moved to the Institute for Advanced Study.[83] But then, life with Robert could not have been easy. Some of the visitors to the Institute realized and appreciated that she did her best to counteract Oppenheimer's coldness and create a friendly atmosphere. She could be caustic, but she was completely free of academic snobbery. She was responsible for starting the Institute's nursery school, which, since its establishment fifty years ago, Freeman Dyson characterizes as "the closest thing we have to a real community." Dyson, who observed her actions during the difficult days of Oppenheimer's trial and his terminal illness, "always saw Kitty as a tower of strength on which Robert leaned heavily."[84]

The period from when he assumed the responsibility for the development of an atomic bomb until his "downfall" was an enormously demanding one on Oppenheimer. I will only adduce one incident as symptomatic of the state of affairs between Robert and Kitty after the war. In the book describing his friendship with Oppenheimer, Chevalier made a point to indicate that when he first met him in 1938, Oppenheimer would never use foul language.

Although they had become close friends, the two of them did not see each other after Oppenheimer left Berkeley for Los Alamos in the spring of 1943 until the war was over. Shortly before Oppenheimer departed, Chevalier informed him that he had been approached by a friend, George Eltenton,[85] who had been told "that Soviet scientists felt that in order to make the most telling use of the scientific know-how and resources of [the United States and the USSR] for the combined war effort it was highly desirable that there be a close collaboration between the scientists of both countries, as there was in other fields." Eltenton knew that Oppenheimer was in charge of an important war project, and in view of his past political sympathies, he thought that Oppenheimer would support the idea of closer scientific coordination and cooperation between the two wartime allies and would be willing to promote the idea. And since Chevalier was a good friend of Oppenheimer, Eltenton had inquired whether he would sound Oppenheimer out as to how he felt about such a collaboration. In his book Chevalier claimed that he immediately saw through the scheme and realized that what the people behind Eltenton were after was knowing more about the secret project that Oppenheimer was working on. Given that perception, Chevalier gave an "unqualified 'No' " to Eltenton's request.[86] Soon thereafter, Chevalier told Oppenheimer of the occurrence of the contact by Eltenton.

Some months later, in August 1943, when the topic of the Federation of Architects, Engineers, Chemists and Technicians (FAECT) came up in an interview with John Lansdale, the security officer at Los Alamos, Oppenheimer decided to tell the security authorities at the Berkeley Radiation Laboratory (Rad Lab) of the Chevalier incident. On a visit to Berkeley in late August 1943, he told the security officer at the Rad Lab that an eye ought to be kept on Eltenton. The security officer in turn contacted Colonel Boris Pash, the person who had jurisdiction over security in the San Francisco area, and reported Oppenheimer's visit. Pash then asked Oppenheimer to see him the next day. At that meeting Oppenheimer told Pash that a friend of his had informed him that Eltenton would be willing to act as an intermediary between scientists at the Rad Lab and the Soviet Consulate. In addition, Oppenheimer said that there had also been a contact with two of his associates at Los Alamos, not by Eltenton but by someone else. But when pressed by Pash, Oppenheimer refused to divulge the name of the person who had effected the contact with the Los Alamos scientists, although he was willing to indicate that the man was a professor. Oppenheimer later admitted that most of what he had told Pash was a "tissue of lies."[87] In December 1943, when urged by General Leslie Groves, Oppenheimer revealed the name of the contact as Haakon Chevalier—who now was being implicated in *three* contacts.

Oppenheimer had been trying to help his friend Chevalier. But by late 1943 Oppenheimer had evidently become "exceedingly ambitious" with regard to his work on the bomb project.[88] By then, he probably considered the USSR a malevolent government. His contact with Groves had made it clear to him that many in the armed forces considered the USSR a potential enemy, and he had accepted the real possibility of Soviet espionage. Oppenheimer was thus caught between his recognition for the needs of security and his sincere efforts to shield his friend. He handled the dilemma clumsily. But the ineptness does not explain why he embellished the story. It might have been an attempt to prove to Groves that he was trustworthy despite his past leftist affiliations. It is an illustration of Isaiah Berlin's perceptive insight into the conflicted emancipated Jews: "They are liable to develop either exaggerated resentments of, or contempt for, the dominant majority, or else over intense admiration or indeed worship for it, or at times, a combination of the two, which leads both to unusual insights and—born of overwrought sensibilities—a neurotic distortion of the facts."[89]

In late May 1946, upon his return to Berkeley after an extended trip to France, Chevalier was interrogated at length by the FBI about his contacts with scientists during the war. He was questioned about his meeting with Oppenheimer and specifically was asked the names of the *three* scientists on the atomic project that he had approached. Shortly after his meeting with the FBI agents, Chevalier met Oppenheimer for the first time since 1943,[90] having been invited to the cocktail party Kitty and Robert were giving at their house to celebrate their return to Berkeley. Chevalier came early to tell Oppenheimer of his recent encounter with the FBI. Oppenheimer's face darkened when Chevalier broached the subject. To secure privacy, they went to the garden, where Chevalier gave Oppenheimer a detailed account of the meeting. Oppenheimer became visibly upset. He told Chevalier that he had reported his approach to him in early 1943 but could offer no explanation as to why Chevalier was being pressed about three contacts. At one stage in the conversation, Kitty asked Oppenheimer to come back into the house as the guests were arriving. Oppenheimer answered "rather abruptly" that he would come in shortly. When Kitty came out a second time to ask him to come in because all the guests had arrived, he "again flung back" a curt answer. When she insisted that he come in, "Opje let loose with a flood of foul language, called Kitty vile names and told her to mind her own goddamn business and to get the . . . hell out."[91]

It is only by taking into account, among many other factors, Oppenheimer's special and charismatic role in the genesis and success of Los Alamos; his transformation from scientist to statesman; his sense of the special responsibility he bore in the development of nuclear weapons;[92] his abhorrence of the plans of the Strategic Air Command and of the activities of its commander, General Curtis LeMay; his passionate efforts during the early stages of the Cold War to bring some sanity into the nuclear arms race;[93] the brutal anticommunism crusade that the government of the United States was waging both domestically[94] and internationally; Oppenheimer's vulnerability by virtue of his deeply felt political activities during the 1930s in support of the cause of the Spanish Loyalist government, of farm workers and teachers, and his loyalty to members of his family with whom he was close, in particular his brother Frank, who had belonged to the Communist party until after the war; the state of his marriage; the nature of the intellectual community at the Institute and of his colleagues there; and the complexity of his personality and character—that we can begin to understand Oppenheimer's conduct in dealing with security matters and, in particular, his behavior in the matter of Bernard Peters.

It is this case and the case of Philip Morrison—and the contrast between Bethe and Oppenheimer in their handling of these matters—that we shall consider in chapter 4. But before doing so, let me outline Oppenheimer's truly outstanding accomplishments as a physicist, as a teacher, and as the builder of an exemplary community during the decade before World War II.

BECOMING A PHYSICIST: OPPENHEIMER AND HIS SCHOOL

During the 1930s Oppenheimer became one of the most influential theorists in the United States, arguably the most imaginative and the deepest of the theorists working on foundational problems. At Berkeley he created "his great school of Theoretical Physics."[95] A very large fraction of the best American theoreticians trained during that decade either obtained their Ph.D. with him or went to Berkeley on postdoctoral fellowships to do research with him: Melba Phillips, Harvey Hall, Leo Nedelski, Frank Carlson, Wendell Furry, Milton Plesset, Arnold Nordsieck, Robert Serber, Fritz Kalckar, Glenn Camp, Stanley Frenkel,

George Volkoff, Julian Schwinger, Hartland Snyder, Bernard Peters, Leonard Schiff, Sidney Dancoff, Philip Morrison, David Bohm, Willis Lamb, Edwin Uehling, Robert Christy, Eldred Nelson, Joseph Keller, William Rarita, Chaim Richman. Oppenheimer and his students probed deeply into the problems of physics at the foundational level, more imaginatively and insightfully than any other school of theoretical physics during the decade. Quantum electrodynamics, meson theory, general relativity, black holes, neutron stars—all came under their critical scrutiny, always with the empirical and the experimentally testable in mind.[96]

Oppenheimer entered Harvard in 1922 with the intention of becoming a mining engineer. He majored in chemistry but took and audited many other courses: in physics, with Edwin Kemble and Percy Bridgman; in mathematics, with George Birkhoff; in philosophy, with Alfred North Whitehead; and in the humanities.[97] "I [there] had a real chance to learn. I loved it. I almost came alive. I took more courses than I was supposed to, lived in the library stack, just raided the place intellectually."[98] In the spring of 1924, he took Physics 6b, a course on advanced thermodynamics taught by Percy Bridgman. In his interview with Thomas Kuhn, Oppenheimer recalled how strongly Bridgman had influenced him:

> I found Bridgman a wonderful teacher because he never really was quite reconciled to things the way they were and he always thought them out; his exercises were a very good way to learn where the bones were in these two beautiful parts of physics. I think that as far as science goes, they were the great point of my time at Harvard. . . . [Bridgman] didn't articulate a philosophic point of view, but he lived it, both in the way he worked in the laboratory which, as you know, was very special, and in the way he taught. He was a man to whom one wanted to be an apprentice.

In many ways, Bridgman embodied the style of thought that Harwood has associated with the American academic culture of the first third of the twentieth century: an empiricist attitude toward problems, a proclivity to choose "simple" problems, and a tendency toward specialization. This in contrast to the German academic culture, which valued generalists tackling "complex" problems.[99]

Oppenheimer's contact with Kemble and Bridgman convinced him that he wanted to be a physicist. But the many hours he spent in Bridgman's laboratory during his senior year "working on a problem on the pressure effect on metallic conduction"[100] had made it clear to him, while acquiring a deep appreciation and understanding of experimental practices, that his "genre, whatever it is, is not experimental science."[101] Nevertheless, upon graduating from Harvard, he went off to Cambridge University to a disastrous experience working in J. J. Thomson's laboratory at the Cavendish Laboratory.

To aspire to be a physicist under the tutelage of Bridgman, and not be very competent in the laboratory, raised doubts about his prospects in the mind of his teachers at Harvard. Bridgman, in his letter of recommendation to Ernest Rutherford at the Cavendish, acknowledged that Oppenheimer's "weakness is on the experimental side. His type of mind is analytical, rather than physical, and he is not at home in the manipulations of the laboratory."[102] Yet it was clear to everyone at Harvard that Oppenheimer was an exceptional person. In his letter to Rutherford, Bridgman also referred to Oppenheimer's perfectly "prodigious power of assimilation." Though he felt unsure whether Oppenheimer would ever make any real contribution of any real character, he indicated that "if he does make good at all, I believe that he will be a very unusual success."[103]

Although J. J. Thomson thought Oppenheimer's experiments "on what happened with beams of electrons and thin films of metals quite good,"[104] Oppenheimer became totally frustrated in the laboratory. His delicate temperament, then already verging on instability, became unbalanced. He recovered, determined to become a theorist. Although he never experienced a crisis of similar proportion, Oppenheimer's delicate emotional balance was seemingly always precarious. As I. I. Rabi later put it: "In Oppenheimer the element of earthiness was feeble."[105] Behind a facade of charm, wit, arrogance, and, on occasion, cruelty also lurked insecurities and doubts about his creativity.

At Cambridge he published his first papers—both on quantum mechanics.[106] They indicate that by mid-1926 he had not only mastered all the recent papers of Max Born, Pascual Jordan, Werner Heisenberg, Paul Dirac, and Wolfgang Pauli, but had gone on to carve out an area of his own: problems involving the continuous spectrum. In the first paper, he showed that frequencies and intensities of molecular band spectra could be obtained unambiguously from the new mechanics. The second paper of Oppenheimer is concerned with the continuous spectrum of the hydrogen atom and discusses the question of the normalization of the wave function in that case.

A visit to Cambridge by Born led to an invitation to continue his studies at Göttingen. Oppenheimer accepted. At Göttingen he became a member of the intellectual community around Born, and blossomed. His close friendship with Dirac and Rabi dates from these days. Edward Condon, who was in Göttingen at the time, recalled that "He [Oppenheimer] and Born became very close friends and saw a great deal of each other, so much so, that Born did not see much of the other theoretical physics students who had come there to work with him. Oppenheimer used to like to work all night and sleep most of the day."[107]

In Göttingen, Oppenheimer continued working on the description of aperiodic phenomena in the new wave mechanics. He was one of the first theorists to work on the quantum-mechanical description of scattering phenomena—problems involving continuum wave functions that could not be tackled with the old quantum theory. One of his papers from that period was devoted to a computation of the absorption coefficient of X rays near the K edge—a very difficult calculation. He also did calculations on the elastic and inelastic scattering of electrons, including one of the first treatments of the exchange effects. With Born, he wrote an important paper that laid the foundation for the quantum-mechanical treatment of molecules, a problem he had already addressed in Cambridge in his paper on the quantum theory of vibrational and rotational degrees of freedom of molecules.[108] In that paper Born and Oppenheimer formulated a quantum-mechanical approach to describing physical phenomena that has become ever more prominent. It can be identified as the first formulation of an "effective" description, an approach that lately has proven very useful and consequential in high energy physics, where it is known under the appellation "effective quantum field theory."[109] Born and Oppenheimer recognized that in molecules the lighter electrons move much faster than the heavier nuclei. To describe the nuclear motion they therefore essentially "integrated out" the (high frequency) electronic motion and obtained an approximate, "effective," wave-mechanical description of the nuclear vibrations.

Oppenheimer obtained his Ph.D. in the spring of 1927 after less than a year's stay in Göttingen. He spent the following year as a National Research Fellow at Harvard and at Cal Tech. At Cal Tech, discussions with Robert Millikan and Charles C. Lauritsen, who had just observed the extraction of electrons from metal surfaces by very strong electric fields, led to an extension of his previous treatment of the ionization of hydrogen atoms by electric fields. His theory of field emission was the first example of the phenomenon of barrier penetration in wave mechanics and antedated Ronald Gurney and Edward Condon's and George Gamow's explanation of alpha decay in radioactive nuclei.[110] As Serber noted in his eulogy of Oppenheimer, his work on cold field emission exhibited a feature that was to become very prominent in his work: "close collaboration with his experimental colleagues."[111] While at Cal Tech, he visited Berkeley and decided that he would like to go there "because it was a desert": "There was no theoretical physics and I thought it would be nice to start something." But he also recognized the danger of being isolated and therefore kept a connection with Cal Tech, which he thought was "a place where I would be checked if I got too far off base and where I would learn of things that might not be adequately reflected in the published literature." He accepted appointments at Berkeley and at Cal Tech that were to start after he had spent another year in Europe as a postdoctoral fellow. He "was a little late in getting to Europe" because he had a slight case of tuberculosis: "It was never very serious but I did go to the mountains for the summer."[112]

In 1929, as a fellow of the International Education Board of the Rockefeller Foundation, Oppenheimer spent half a year with Paul Ehrenfest and Hans Kramers in Holland. Although he had planned to go to Copenhagen to work with Niels Bohr, "Ehrenfest's certainty that Bohr with his largeness and vagueness was not the medicine [he] needed but that [he] needed someone who was a professional calculating physicist and that Pauli would be right for [him]"[113] brought Oppenheimer to the Eidgenössische Technische Hochschule (ETH) in Zurich. Pauli was by then devoting almost all of his efforts to the problems of relativistic quantum mechanics and to the quantum-mechanical description of the interaction of charged matter with the electromagnetic field. When Oppenheimer got to Zurich, Pauli told him of his work with Heisenberg on the quantum theory of fields, and Oppenheimer showed "more than a little interest in it."[114] He got deeply involved in these problems and made important contributions to them, clarifying among other things the relativistic invariance of the quantization rules that Heisenberg and Pauli had devised for fields, and the configuration space treatment of theories in which particles (such as photons) are created and destroyed.

> At first [we] thought the three of us should publish together; then Pauli thought he might publish it with me and then it seemed better to make some reference to it in their paper and let this be a separate publication. But Pauli said, "You really made a terrible mess of the continuous spectra and you have a duty to clean it up, and besides, if you clean it up, you may please the astronomers."[115]

Oppenheimer's interest shifted to these more fundamental questions during his stay in Zurich. Upon his return to Berkeley, he devoted himself to understanding every facet of quantum electrodynamics: its experimental predictions, its range of validity, and the internal difficulties that beset it. By the fall of 1929, Oppenheimer had completed an important study of the self-energy problem of an electron,[116] basing himself on the formulation of quantum electrodynamics that Heisenberg and Pauli had given.[117] He discovered that the energy levels for a bound electron in a hydrogen atom were shifted by an infinite amount due to the interaction of the electron with the field it created. This level shift was quadratically divergent because Oppenheimer had made the calculation on the assumption that all the negative energy states in the description of an electron by the Dirac equation[118] were unoccupied. Nonetheless, he hoped that observable effects such as the frequency of the radiation emitted in a transition between two states might be finite because the frequency is calculated by taking the difference of the energies of two states.[119] He observed that the leading divergent terms were equal for states of the same energy, and found that the expression for the frequency of the emitted radiation "diverges logarithmically for high frequencies." Had this calculation been repeated using positron theory—which could certainly have been done any time after 1932, the "Lamb" shift[120] would have been predicted theoretically, for in a hole-theoretic calculation, the level shifts are logarithmically divergent and the energy difference between two levels is finite.

In 1930, Dirac made his famous suggestion on how to overcome the problem of the negative energy solutions of the spin relativistic wave equation that he had advanced in 1928 to describe an electron.[121] He postulated that there are so many electrons in the world "that all the states of negative energy are occupied except a few of small velocity. . . . Only the small departure from exact uniformity, brought about by the negative energy states being unoccupied, can we hope to observe." He further assumed that the holes in the distribution of negative energy electrons were protons. It is interesting to compare Pauli's and Oppenheimer's objections to this identification of the protons with the holes. Pauli "rigorously" proved that Dirac's Hamiltonian—which included the Coulomb interaction of the charged particles with one another and their interaction with the quantized radiation field—"implied that the system consisting of m positive energy electrons and n 'holes' . . . has the same energy as the system consisting of m holes and n electrons, the electrons having the velocities which previously belonged to the holes and vice versa."[122] Hence holes have the same mass as electrons and cannot be identified with protons. Oppenheimer's proof that the positively charged particles of the Dirac theory must have the same mass as the electrons was based on a physical argument.[123] He showed that the *experimentally* established Thomson limit for the scattering of low energy photons off electrons could not be recovered from the Klein-Nishina formula for Compton scattering if the electron mass was used for the positive energy states and the proton mass for the negative energy states. Oppenheimer also calculated the lifetime for annihilation of a particle and an antiparticle. The similarity between the matrix elements involved in this calculation and those for Thomson scattering was adduced as an argument for believing the lifetime for annihilation that he had computed. Since a hydrogen atom in its ground state is stable and does not decay, Oppenheimer concluded that the proton must be an *independent* spin ½ elementary particle with its own antiparticle!

When Dirac advanced his hole picture, he was aware that the theory was symmetrical between positive and negative energies, so that the hole should have the same mass as the electron. In 1978, Dirac recalled that

at that time the only positively charged particle that was known was the proton. People believed that the whole of matter was to be explained in terms of electrons and protons, just those two particles. One needed only two particles because there were only two kinds of electricity, negative and positive. There had to be electrons for the negative electricity and protons for the positive electricity, and that was all. I just didn't dare to postulate a new particle at that stage, because the whole climate of opinion at that time was against new particles. So I thought this hole would have to be a proton. I was very well aware that there was an enormous mass difference between the proton and the electron but I thought that in some way the Coulomb force between the electrons in the sea might lead to the appearance of a different rest mass for the proton. So I published my paper on this subject as a theory of electrons and protons.[124]

But the objections by Oppenheimer[125]—and by Tamm,[126] who also had computed the rate of annihilation of protons and electrons into gamma rays on the basis of Dirac's hole theory (with holes corresponding to protons) and had found that under these assumptions the hydrogen atom would be unstable—and Weyl's proof that left-right, mirror inversion symmetry demanded that a hole have the same mass as the electron forced Dirac to propose that "a hole, if there were one, would be a new kind of particle, unknown to experimental physics, having the same mass and opposite charge to an electron. We may call such a particle an anti-electron."[127] In his talk before the British Association for the Advancement of Science (BAAS) in Bristol in 1930, Dirac indicated that Oppenheimer had suggested that "all and not merely nearly all" of the negative energy states for an electron are occupied, so that a positive energy electron can never make a transition to a negative energy state.[128]

Patrick Blackett—who was working at the Cavendish in Cambridge at the time—told Dirac that he and Occhialini had evidence for this new kind of particle.[129] But Blackett did not want to publish his results without further proof, and "while he was obtaining the corroboration, Carl Anderson quite independently published his evidence to show that the positron really existed."

The history of the discovery of the positron provides a vivid illustration of the struggles to overcome ingrained ways of seeing the world. What is striking is the conservatism of some of the leading theorists—such as Dirac, Pauli, and Bohr—with regard to the building blocks of matter: They were reluctant to assume the existence of new particles, although they endorsed revolutionary stands in the formulation of theories.[130] Oppenheimer, on the other hand, was open to both possibilities.

In 1933, Oppenheimer and Milton Spinoza Plesset calculated the cross section for pair production by gamma rays in the Coulomb field of a nucleus, and showed that hole theory quantitatively explained the excess absorption of ThC' gamma rays in heavy elements. But they also came to the conclusion that while the theory could account for phenomena in the energy range characterized by the electron mass m (energies of the order mc^2, i.e., a few Mev's or less) the theory must fail at energies greater than $137\ mc^2$. During the 1930s, Oppenheimer, like most of the other leading theorists, believed that quantum electrodynamics was incorrect, a view he repeatedly stressed.[131] This belief may have been "a fundamental barrier to Oppenheimer's success in making progress with the difficulties of quantum electrodynamics."[132]

The calculations Oppenheimer undertook were long, involved, and very complicated. Although everyone admired the physical content of his papers, the accuracy of his results was taken somewhat skeptically, for he was well known to make numerical mistakes.[133] In 1933, when Harvard was trying to get Oppenheimer to join its faculty, Ralph Fowler[134] wrote a qualified recommendation: "I fancy he is not a very good lecturer and his work is still apt to be full of mistakes due to lack of care, but it is work of the highest originality and he has an extremely stimulating influence in a theoretical school as I had ample opportunities of learning last fall."[135]

The papers that Oppenheimer and his students wrote during the 1930s were produced with little direct contact with the rest of the theoretical physics community. They only read the published literature. Both Furry, who was a postdoctoral fellow with Oppenheimer from 1932 to 1934, and Serber, who was at Berkeley from 1934 to 1938 in a similar capacity, have commented on the isolation of the Berkeley theory group: "We were too poor to travel."[136] At times, Oppenheimer and Ernest Lawrence made trips to the East Coast to attend meetings of the American Physical Society and, when back in Berkeley, would report on the new physics they had learned. There were also occasional visitors: Ralph Fowler in 1932, Niels Bohr in 1933, Enrico Fermi in 1935. But on the whole, it was Oppenheimer who dominated the intellectual scene and determined the research projects. As Bethe noted in his deeply felt eulogy of Oppenheimer: "His teaching, his style and his example formed the scientific attitude of all [his students]."[137]

Wendell Furry vividly remembered the impression "Oppie" made on him when he came to Berkeley in 1932 after completing his Ph.D. with James Bartlett at the University of Illinois. "Oppenheimer bowled me over. I was very depressed soon after beginning and hearing him talk. He was not very patient."[138] Furry spent his first year at Berkeley learning field theory from the papers of Pauli and Heisenberg, and those of Fermi[139] and Gregory Breit,[140] since "the latter were much more understandable." He also became involved in the research program that Oppenheimer had initiated to investigate the various processes responsible for the attenuation of high energy particles in their passage through matter.

In the spring of 1933, Niels Bohr visited Berkeley and lectured on his work with Leon Rosenfeld on the problems of the measurability of the electromagnetic field.[141] Furry recalled that these issues "were taken very seriously" and stressed that Bohr's visit was "highly valued" as a window on the world of Pauli, Heisenberg, Weisskopf, and the other European theoreticians working on the problems of quantum electrodynamics. This elite circle corresponded extensively among themselves[142] and congregated annually in Copenhagen for the presentation and discussion of ongoing work. At Bohr's suggestion, Franklin Carlson and Furry computed the self-energy of an electron in hole theory. However, they calculated only the magnetic self-energy, believing that the electrostatic self-energy would still diverge linearly, as would be expected from classical theory, and would be unaffected by hole theory. They obtained a logarithmically divergent expression—an important result since the magnetic self-energy of an electron diverges quadratically when calculated in the one-particle Dirac theory (in which transitions to negative energy states are possible).[143] Early in 1934, Weisskopf, who at the time was Pauli's assistant, undertook a similar calculation at Pauli's request. His analysis was complete and included the electrostatic proper energy "which for some reason [Carlson and Furry] had not realized the need of recalculating."[144] Weisskopf found a logarithmically divergent expression for this part of the self-energy. However, he made an error in his computation of the magnetic self-energy and obtained a linearly divergent result for that contribution. When Weisskopf's paper appeared in the *Zeitschrift für Physik*,[145] Furry and Carlson read it and noted the error in the magnetic term. They went to Oppenheimer and asked him what to do. He indicated, "You can either publish or do the noble thing."[146] The echo of Adler's teachings at the Ethical Cultural School in Oppenheimer's admonition should be noted. Furry and Carlson did the "noble thing." Furry wrote Weisskopf informing him of the result of their previous calculation of the magnetic self-energy and indicated to him where he had made his error.[147] Weisskopf proceeded to publish a note correcting his previous calculation of the self-energy, acknowledging Furry and Carlson's intervention.[148]

When Oppenheimer first came to Berkeley and started giving graduate courses, Fowler's guess regarding his teaching was correct: his lectures did leave students bewildered. But after a few years, all his courses—particularly his lectures on quantum mechanics and on electromagnetic theory—were models of clarity, emphasizing utility but also conveying to the students the beauty of the subject matter. Bethe noted: "Probably the most important ingredient he brought to his teaching was his exquisite taste. He always knew what were the important problems, as shown by his choice of subjects. He truly lived with these problems, struggling for a solution, and he communicated his problems to his group."[149] In his eulogy, Serber described Oppenheimer's interaction with his students:

> He met the group [which in the mid-thirties consisted of a dozen graduate students and about half a dozen postdoctoral fellows] once a day in his office. A little before the appointed time the members straggled in and disposed themselves on the tables and about the walls. Oppie came in and discussed with one after another the status of the student's research problem while others listened and offered comments. All were exposed to a broad range of topics. Oppenheimer was interested in everything; one subject after another was introduced and coexisted with all the others. In an afternoon they might discuss electrodynamics, cosmic rays, astrophysics, and nuclear physics.[150]

This form of cooperative investigation that made use of the collective knowledge of the group became the characteristic mode of distributed inquiry in research groups—but only much later. In his autobiography, Serber added some further observations: "As Oppie finished with each student he would advise him how to continue. Finally, he would leave. Most of the crew would stay, and my job began: to explain to each what Oppie had told him to do. They were much more willing to display lack of understanding to me than they were to Oppie."[151]

Oppenheimer was not only the center of his students' intellectual world, he was also at the center of their social world. Oppenheimer and his students frequently went to one another's houses for parties, for dinner, or to play bridge; they also often went to the movies together.[152] It became a custom for everyone to go out to dinner after the Friday afternoon theoretical physics seminar, which alternated between Berkeley and Palo Alto. This weekly colloquium had been instituted after Felix Bloch's arrival at Stanford in 1933. Oppenheimer almost always paid the bill for the dinner.[153]

Oppenheimer's researches during the 1930s dealt primarily with subjects at the frontiers of physics: cosmic rays, quantum electrodynamics, nuclear physics, and astrophysics. All his investigations and those of his students—no matter how esoteric or formal—always had the experimental situation in mind.[154] Often they were motivated by experimental findings at Cal Tech and at Berkeley. Outstanding examples are the attempt by his student, Edwin Uehling, to explain the deviations in the spectrum of hydrogen from the predictions of the Dirac equation as a result of vacuum polarization;[155] his theory of pair production; and his explanation of cosmic ray showers as being initiated by electrons emitting X rays, which X rays in turn produce electron-positron pairs in the electric field of the atomic nuclei of the atmosphere.[156] In the summer of 1936, Oppenheimer had discussed in a qualitative manner his theory of electron-positron showers, and by the end of that year he and Carlson gave an elegant treatment of this process. The success of shower theory corroborated for Oppenheimer the validity of quantum electrodynamics at low energies, and from the discrepancy between the theory and the empirical data he inferred the existence of a new type of particle in cosmic rays.

After the discovery of the neutron, nuclear physics became one of Oppenheimer's main interests.[157] He and Ernest Lawrence were close friends,[158] and Oppenheimer followed closely the researches of the cyclotron group.[159] Experimentalists and theorists met regularly at the weekly Wednesday afternoon colloquia and the Tuesday evening Journal Club. The Journal Club, at which the most recent theoretical and experimental papers were presented and discussed, was designed as an informal vehicle for exchanges of information between theorists and experimenters. Oppenheimer dominated the scene here also. Milton Livingston, who was Lawrence's assistant during the early 1930s and helped him build the first cyclotron, recalled that

Lawrence won "general admiration" by not being fazed by Oppenheimer. "He didn't give a damn if he asked foolish questions. At seminars and Journal Club we [the experimentalists] sat afraid to ask Oppenheimer anything. Lawrence would pop up and ask something silly.[160] Robert Wilson remembers Lawrence dragging him to the Journal Club because even though he wouldn't understand anything it would be an intellectual experience watching Oppenheimer in action.[161]

Oppenheimer overwhelmed the experimentalists not only in theory. Raymond Birge, the chairman of the physics department at Berkeley, recalled that "in our seminars Oppenheimer knew more experimental physics than even the experimental physicists did. He could reel off figures and equations relating to experiments better than any experimental physicist in the room."[162] When Walter Elsasser met Oppenheimer in 1936, he was struck by the fact that "Oppenheimer knew by heart every one of the many hundreds of nuclear reactions that had been discovered by then; he not only remembered them almost instantly; he also remembered the quantitative details and the place where they had been published."[163] At the memorial session for Oppenheimer at the 1967 Washington meeting of the American Physical Society, Robert Serber spoke about Oppie's Berkeley years. He sent a reprint of his talk to Leslie Groves, the general who had been in charge of the Manhattan Project and had picked Oppenheimer as director of Los Alamos despite his ambiguous security file. In his reply, Groves wrote: "I was pleased with the emphasis you placed on his ability to work in close collaboration with the experimentalists and his most unusual ability to inspire his students and his sensitive perception of their problems. These outstanding characteristics were the basis for my selection of him for the Los Alamos post which he filled so admirably"[164]

In 1935, Hideki Yukawa published a paper in which he proposed a field theoretical model to account for the nuclear forces. In Yukawa's theory, the neutron-proton force was mediated by the exchange of a scalar particle between the neutron and proton; the mass of the scalar particle—later called a meson—was adjusted to yield a reasonable range for the nuclear forces. Yukawa had writ large what had been known in QED, namely that the electromagnetic force between charged particles could be conceptualized as arising from the exchange of (virtual) photons—virtual because these photons did not obey the relation $E = h\nu$ valid for free photons.[165] In June 1937, immediately after Seth Anderson and Carl Neddermeyer had given evidence for the existence of a new type of particle in the penetrating component of cosmic rays, Oppenheimer and Serber published a short note in the *Physical Review* pointing out that the mass of the newly discovered particle specified a length which they connected with the range of the nuclear forces, as had been suggested by Yukawa. Oppenheimer and Serber's note was responsible for drawing the attention of American physicists to the "meson theories" of nuclear forces that Yukawa had advanced. The existence of this "heavy electron"—which existed in both a positive and a negative variety— was authenticated by its direct observation in a cloud chamber by Curry Street and C. E. Stevenson.[166] Oppenheimer and his students thereafter became deeply involved in devising empirical tests to determine the spin of these particles and in the elaboration of meson theories.

Oppenheimer's yearly trek to Pasadena led to a deep and rewarding friendship with Richard Tolman and to influential interactions with the staffs of Mt. Wilson Observatory and of Palomar, in particular with Walter Baade and Fritz Zwicky. His discussions with Tolman, and later with Baade and Zwicky, were responsible for his important explorations in astrophysics and his investigations of general relativity.[167]

In 1932 Lev Landau had given a general argument indicating that a sufficiently large agglomeration of cold matter could not sustain itself against gravitational collapse. His argument was based on the idea that electrons, as particles obeying Fermi-Dirac statistics, would supply the pressure resisting collapse. As the mass of the aggregate increases, the material becomes more and more compacted, and as a result the energy per particle of the electron Fermi gas increases to higher and higher values and eventually becomes relativistic, whereupon it ceases to depend upon the rest mass. It then makes no difference in the energy per particle, and hence in the pressure, whether the Fermi gas is made up of electrons or neutrons. Both the compressional and the gravitational energy then depend the same way on the effective radius of the assembly, namely as $1/R$. The key issue is then the sign of the total energy per particle. When it is negative, collapse will set in. Landau arrived at a mass of the order of that of the sun as the critical mass beyond which collapse occurs. In fact, already in 1931 Subrahmanyan Chandrasekhar had made a more accurate calculation than Lev Landau, and found that, for masses greater than 1.4 times the mass of the sun, no equilibrium can exist: the electron pressure is always overwhelmed and such a star cannot support itself against gravity.

In 1938, Landau wrote an article that was published in *Nature* in which he suggested that massive stars might contain a neutron core at their center. The article drew Oppenheimer's attention. Oppenheimer and Serber pointed out that if such a core existed at the sun's center and were responsible for a large fraction of the sun's mass, the sun's radius would be much smaller than its observed value. Moreover, by taking more fully into account the nuclear forces between neutrons, they came to the conclusion that Landau's estimate of 0.001 solar mass for the minimum mass of a neutron core was wrong. The nuclear forces precluded the existence of a neutron core for stars with masses comparable to that of the sun.

Landau's suggestion of neutron cores within stars drew Oppenheimer to a reconsideration of Zwicky's idea of neutron stars. Bethe and Critchfield had thoroughly investigated the mechanism for energy generation in stars due to the $p + p \rightarrow d + e^+ + \upsilon$ reaction and had published their results in the *Physical Review*.

The fate of stars once they exhausted their nuclear fuel was the problem to be addressed next. Would a massive star degenerate into a white dwarf? Chandrasekhar had shown unequivocally that stars less massive than 1.4 suns evolve into white dwarfs. What about stars heavier than 1.4 suns? Zwicky had suggested that such stars would implode into neutron stars, and that such an implosion was related to the creation of supernovas.

Oppenheimer and Volkoff analyzed numerically the equilibrium of a star built up from an ideal neutron gas.[168] For this purpose they employed for the first time the general-relativistic equation for hydrostatic equilibrium, and found that stable equilibrium can only exist for masses smaller than 0.7 times the mass of the sun. In a subsequent paper, Oppenheimer and Hartland Snyder studied the "indefinite" collapse of a heavy star when all thermonuclear sources of energy have been exhausted by analyzing the solutions of the gravitational field equations that describe this process.[169] In particular, they gave general and qualitative arguments on the behavior of the metric tensor as the contraction progresses: the radius of the star asymptotically approaches its Schwarzschild radius, and light from the surface of the star is progressively reddened and able to escape only over a progressively narrower range of angles. In addition, for the case that the pressure in the star is neglected, they obtained an analytic solution of the field equations confirming their general argument. They had described what John Archibald Wheeler later called a black hole.

Oppenheimer's letters to Bethe, Pauli, Uhlenbeck, and Wigner during the 1930s attest to his mastery of all aspects of nuclear and "high energy" physics, ranging from a detailed knowledge of nuclear energy levels, nuclear reaction cross sections, beta-ray spectra, to the most recondite calculations in meson theory, beta-decay theory, and quantum electrodynamics.[170] In 1954, Oppenheimer reviewed his activities at Berkeley in the prewar period:

Starting with a single graduate student in my first year in Berkeley, we gradually began to build up what was to become the largest school in the country of graduate and postdoctoral study in theoretical physics, so that as time went on, we came to have between a dozen and 20 people learning and adding to quantum theory, nuclear physics, relativity and other modern physics. As the number of students increased, so in general did their quality. The men who worked with me during those years hold chairs in many of the great centers of physics in the United States; they have made important contributions to science.[171]

His assessment was accurate.

3.

HANS BETHE

"Make of yourself a light,"
said the Buddha, before he died.
I think of this every morning
as the east begins
to tear off its many clouds
of darkness . . .
—Mary Oliver, "The Buddha's
Last Instruction" (1992)

BECOMING A *BILDUNGSTRÄGER*

Bethe was born on July 2, 1906, in Strasbourg, when Alsace was part of the Wilhelminian empire. He is an only child. His father was a widely respected physiologist who accepted a professorship in Frankfurt when Hans was nine years old. His mother was raised in Strasbourg where her father had been a professor of medicine. Bethe's father was Protestant. His mother had been Jewish but became a Lutheran before she met Hans's father at the ceremony at which her older sister converted in order to marry a German army officer. Hans thus grew up in a Protestant household, but one in which religion did not play an important role. His mother was a talented and accomplished musician, but a year or two before World War I her hearing was impaired as a result of contracting influenza. The illness affected not only her sense of pitch and put an end to her public performances as a singer, but it also left psychological scars. Everyone who knew Hans's mother remembers her as very difficult; she was prone to what was diagnosed at the time as bouts of "nervous exhaustion," extended periods of depression.

The high school Bethe attended in Frankfurt was a traditional *humanistisches Gymnasium* having a heavy emphasis on Greek and Latin. Although instruction in mathematics and the sciences at gymnasia was in general poorer than in the humanities, several of the mathematics instructors at the Goethe Gymnasium were quite competent. Mr. Wirtz, who was one of Hans's math teachers, was both well qualified and stimulating, and Hans mastered a fair amount of mathematics under his tutelage. At the Gymnasium Hans learned Latin and Greek, read Kant, Goethe, and Schiller, and also learned

French and English and a good deal of science. Classes were from 8 A.M. until 1 P.M., six days a week,[1] with lots of homework assigned daily. There were approximately thirty to thirty-five students in the class, roughly one half Protestant, one half Jewish, and one Catholic. All the teachers were men and all of them were of the Christian faith.

Karl Guggenheim was in the same class in the Goethe Gymnasium as Hans from 1918 until 1924. He remembers Hans as having been among the most gifted in the class in every subject and to have particularly "distinguished himself in mathematics and physics." Hans also stood out by being very helpful to the other boys in these classes. Guggenheim recalls Hans being slightly heavy set and poor in gymnastics, and that he was struck by the incongruity of Hans being so good in the classroom and so uncoordinated in gym.[2] Another student in the class was Felix Goldschmidt, a close friend of Karl Guggenheim who emigrated to Palestine in 1933.[3] Guggenheim remembers Goldschmidt working in the phosphate works on the southern tip of the Dead Sea, reciting with much pleasure Homer in Greek and Horace and Ovid in Latin.

The Goethe Gymnasium groomed Bethe to be a *Bildungsträger*.[4] Central to the concept of *Bildungsträger* was the notion of *Bildung*. *Bildung* fuses the meaning we attach to the word "education" with beliefs about character formation and moral education aimed at the development of the whole person. The acquisition of *Bildung* came to be seen as a continuous, never-ending process in which reason and esthetic sensibilities were cultivated: "Its purpose was to lead the individual from superstition to enlightenment."[5] *Bildung* became a central tenet in the ideology of the German professorate during the nineteenth century; *Bildungsideal* was the dogma of the Gymnasiums and universities in contrast to the *Berufsideal* (professional goal) of the *Realschule* and *Technische Hochschule*.

From 1790 to 1810, Wilhelm von Humboldt was an important contributor to the development of the *Bildungsideal* and its associated concepts of self-cultivation and self-formation. As the government official in charge of educational reform in Prussia in 1809–10, he was responsible for the canonization of *Bildung* by enshrining it in the articles of governance of the University of Berlin. He also deeply influenced the course of Jewish emancipation during the first two decades of the nineteenth century by virtue of a memorandum he wrote to the chancellor of Prussia in 1809 and of his activities at the Council of Vienna.

Just as Felix Adler loomed large in Oppenheimer's cultural background, so did Wilhelm von Humboldt (1767–1835) in Bethe's. A brief sketch of his personal biography is thus in order. I should add that there is also a striking similarity between Wilhelm von Humboldt's character and personality and that of Bethe's. Both share the trait that they could bare their soul only to women, and that their closest friends were women—emancipated German Jewish women.

Wilhelm von Humboldt's father was an army officer who became a *Kammerherr*, a courtier-counsellor, in the household of the Prussian crown prince, the future Frederick William II. His mother came from a Huguenot family that had settled in Berlin in the early 1700s. A previous marriage had left her widowed, as a result of which she inherited a considerable fortune and a sizable estate in Tegel. She married Major von Humboldt in 1766; Wilhelm was born in 1767, Alexander two years later. Wilhelm's father died when he was twelve, and both he and Alexander spoke of the home as being an unhappy place after the death. The mother was distant, and G.J.C. Kunth, the tutor she had hired to oversee the education of the two sons to train them for a civilian career in the Prussian government, was a demanding and domineering individual. Wilhelm learned to hide his emotions and to cultivate skills in dissimulation in order to satisfy his tutor's wishes. Although hypersensitive and introverted, he learned to project an image of affability, and also acquired the talent to amuse. Finding the atmosphere in the house oppressive, he "felt a compulsion to distance himself from what went on around him" and developed an "overwhelming" need for self-control.[6] The serenity he projected was achieved through self-mastery and acquired at some cost. Throughout his life he cultivated two qualities that shaped his personality from his youth on: will-power and self-control.

By 1785, both Wilhelm and Alexander were spending most of their time in Berlin. Their private teachers there belonged to the Lessing-Mendelsohn-Nicolai circles,[7] and through them they came into contact with the wealthy, educated Jews of Berlin—in particular, with Moses Mendelsohn, David Friedländer, Marcus Herz and Simon Veit[8]—and began frequenting the *salons* of Henriette and Marcus Herz, Brendel Veit, and Rahel Varhagen. Friedrich Schleiermacher, who was to enter this same circle a few years later, explained why the salons of the Herzes, Verhagen, and other rich Berlin Jews attracted young scholars. He noted that these Jewish households were

> by far the richest bürgerlichen [middle-class] families here, and almost the only ones who hold an open house. . . . Whoever wants good society of an informal sort gets introduced to such

houses and is certain to be amused because the Jewish women (the men are plunged too soon into business) are well educated, can talk about everything and usually are well versed in one or the other of the arts.[9]

These salons—animated by the emancipatory promise and hope of the *Aufklärung* to Jews and women—created an environment in which the seeds of the revolutionary changes that were taking place were cultivated. In fact, in these salons the promise was fulfilled and acted out. These Jewish *salonières* enthusiastically supported the young advocates of the Romantic *Sturm und Drang* movement and were influential in the German rejection of French classicism in the arts and in literature at the beginning of the nineteenth century.[10]

Wilhelm became very close to Henriette Herz, whose husband Marcus was one of Kant's first students and the leading exponent of critical philosophy in Berlin. Wilhelm bared his soul to her and became a member of the *Tugendbund* that Henriette had organized in 1787. This "league of virtue" was a secret society dedicated to mutual moral improvement and to the development of a deeper knowledge of humanity. Promotion of the love of one's fellow man was the Tugendbund's avowed purpose. Of the six initial members of the Tugendbund, three were women: Henriette Herz, Brendel (Dorothy) Mendelssohn (who became Dorothy Veit and later the wife of Wilhelm von Schlegel), and Caroline von Dacheröden, who later married Wilhelm. Henriette taught Wilhelm Hebrew and Yiddish, and a number of his letters to her and to his friend Beer from the Tugendbund were written using the Hebrew alphabet as in Yiddish. Their relationship cooled after he went to Göttingen in the spring of 1788. His circle of friends there included Israel Stieglitz, a Jewish medical student, whom he had met in Henriette Herz's salon and who became his best friend.[11]

In Göttingen, Humboldt immersed himself in lectures at the university and on his own started an intensive study of Leibniz's works and of Kant's *Critique of Pure Reason*. Humboldt had been introduced to Leibniz's philosophy and to his beneficent worldview by attending the lectures of J. J. Engel the at University of Berlin. Humboldt's notion of *Bildung* relied on a very particular image of man that he had gotten from his intensive reading of Leibniz during 1788–89. In his *Limits of State Action*, Humboldt stated "Energy (*Kraft*) appears to me to be the first and unique virtue of mankind."[12] An individual's development hinges on finding appropriate conduits to channel his energy, that is, activities and forms of expression by means of which he realizes his potentialities and increases his abilities. An essential condition for this is freedom. One must have the freedom to act for oneself, that is, freedom to be self-reliant. Humboldt constantly stressed that *Bildung* required individual freedom, and that the unfolding of human potentiality to its maximum could not brook any interference. He believed that all institutions that hinder *Bildung* were harmful. He also advanced a second condition for the development of the whole person: "social intercourse." An individual unfolds through the voluntary interchange of his individuality with that of others.[13] Humboldt had learned from the Tugendbund that self-formation requires social bonds, that is, a network of personal relationships.

The political and social harmony that *Bildung* bestowed had to be achieved in the modern state. It had existed in ancient Greece where person and citizen were one. For Humboldt, as for the Greeks, citizenship and individuality were to be indistinguishable, but with the understanding that individuality implied that citizenship could not exist without individual freedom.

Humboldt's theory of *Bildung* required a concept of harmony analogous to the one Leibniz had promulgated, for *Bildung* required interactions between individuals who enjoyed unfettered freedom. Leibniz had predicated his concept of individual monads of energy on the metaphysical presupposition of a "preestablished harmony" of the universe. Only with this concept of harmony could Leibniz maintain a notion of a unified world because his individual monads had no "windows."

As elaborated by Humboldt, esthetics was the foundation for the harmony that *Bildung* imparted on an individual. The *gebildete* person was compared to a harmonious work of art.

> The aesthetic was linked to the intellectual faculty, and both activated the moral imperative that resided in man.
>
> The beautiful as the essence of aesthetic education was not romanticized but understood through reason. . . . The beautiful, in accordance with the Greek ideal, was conceived as harmonious and well proportioned, without any excess or false note which might upset its quiet greatness. Beauty was supposed to aid in controlling the passions, not in unleashing them, emphasizing that self-control which the bourgeois prized so highly. The ideal of Greek beauty transcended the daily, the momentary; for Humboldt, as for Goethe and Schiller, it symbolized the ideal of a shared humanity towards which *Bildung* must strive. This beauty was a moral beauty through its strictness and harmony of form; for Schiller it was supposed to keep humanity from going astray in cruelty, slackness, and perversity. *Bildung* was not chaotic or experimental but disciplined and self-controlled"[14]

With the attainment of the harmonious personality came enlarged perspective and balanced judgment that allowed a person to apprehend the whole truth—indeed, to obtain "a living picture of the world properly unified."[15]

Bildung became emblematic of the German educated middle class during the latter part of the eighteenth century. To be a *Bildungsträger* implied membership in both class and nation. The writings of Lessing, Herder, Goethe, and Schiller gave *Bildungsträger* their sense of cultural identity. *Kulturträger* saw themselves as expressing intellectually the unity of the German-speaking states—prefiguring their eventual political unification. The concept of *Bildung* acquired a more nationalistic and political tone following the French military victories at the beginning of the nineteenth century. The educated middle classes clamored for reforms that would give them a greater political voice and demanded that the state support *Bildung* in the Gymnasiums and universities. Prussia was the first to acquiesce, recognizing that educational reforms would not only secure middle-class allegiance, but also provide the state with loyal civil servants.

As head of the section for religion and education in the ministry of education of Prussia from February 1809 to June 1810, Humboldt formulated the guidelines that would make *Bildung* the objective not only of the newly founded University of Berlin but of the entire Prussian educational system. In Humboldt's program, the state was to assume financial responsibility for education throughout the nation. He envisioned that in elementary school, students would learn basic skills; in high school they would be taught to be intellectually independent. A student was to be considered mature when "he had learned enough from others to be able to learn by himself." The main function of the university was to join students and professors in an elite academic community based solely upon mutual self-cultivation through learning (*Wissenschaft*). University professors would be free to teach what they desired (*Lehrfreiheit*), and students would be free to pursue the course of study they chose (*Lernfreiheit*)—without the state interfering in any way.[16] By congregating students in a community steeped in Wissenschaft and giving them freedom to interact with their peers, Humboldt hoped they would acquire models to emulate and cultivate. By midcentury, *Bildung* was institutionalized in the Gymnasium and the universities and overseen by a new mandarinate: the university professors, and much lower in the pecking order, the *Oberlehrer*, the Gymnasium professors.

Despite Humboldt's influential reconciliation of *Aufklärung* and neohumanist conceptions of *Bildung*, and despite the genuinely egalitarian aspects of his concept of general education, his educational reform program came to serve functions diametrically opposed to those for which it had been designed. Institutionalized *Bildung* became the social and ideological basis for a self-perpetuating elite who lacked "the internal imperative of political activities" and was "wedded to the state in exchange for the state's identification with culture." Ringer called its members "mandarins."[17]

The universities did maintain academic freedom—*Lehrfreiheit*—and this freedom and the ideals of *Bildung* did produce impressive results: they propelled German scholarship to the forefront of the humanities, mathematics, and the sciences. But after German unification, the professorate, claiming to be the conscience of the nation, became the guardians of the status quo. With Prussia setting the example, patriotism, duty, and discipline threatened to replace skepticism, critical inquiry, and openness, ideals fostered by the Enlightenment. The social and economic changes since the 1850s, brought about by industrialization and urbanization and by the legislation nurturing the democratization of Germany, had begun, by the last two decades of the century, to undermine the privileged position of the educated middle classes. They became more conservative—and the professorate reflected these structural changes.

Emancipated German Jews were an important component of this educated middle class. The French Revolution had initiated the formal, legal emancipation of the Jews of Europe. The Constituent Assembly had given French Jews full citizenship and equal rights. These prerogatives were brought to Germany by a victorious Napoleon and had been declared the law of the land in the conquered territories.

Wilhelm von Humboldt's contacts with the cosmopolitan, emancipated, wealthy German Jews of Berlin, his friendship with Henriette Herz and the other Jewish members of the *Tugendbund,* and with Stieglitz in Göttingen had made him favorably disposed to the granting of complete rights of citizenship to the Jews of Germany.[18] In 1809, Humboldt prepared a paper recommending equal political rights for Jews. The subsequent Prussian Edict of Emancipation of 1812 did in fact grant Jews almost civil equality and made it possible for them to enter the Prussian educational institutions. Although Jews could not become civil servants unless they became Christian, and hence could not become ordinary professors at Prussian universities without converting, they could now attend Gymnasiums and universities and become doctors and lawyers. Many of them did.

In his 1809 memorandum, Humboldt had emphasized the need for uniform action throughout all the German states. His motivation was a combination of idealism and realism. As a modern nation, Prussia should be committed to the emancipation of the Jews. It needed to consolidate its power and have sovereignty over all its subjects without the mediation of social or religious institutions, and therefore could not tolerate the autonomous corporate status of the Jewish community. Uniformity of practice in all the German states was required "so that the Jews would not be lured into Prussia by the circumstance that the Jews enjoy much greater privileges there than in the rest of Germany."[19] Humboldt regarded Jews not as a people but as a multitude of individuals, and equality was to lead to the disappearance of the Jews as an ethnic group. This became the official position. In return for Jewish emancipation, Germans Jews would have to disavow Jewish peoplehood—a price many secularized Jews were willing to pay.[20] But the price of emancipation—becoming autonomous individuals in a bourgeois society—raised the problem of what it meant to be Jewish. The answer was provided by *Bildung*.

For secularized German Jews, *Bildung* and Enlightenment became melded.[21] They interpreted the message of the Enlightenment as attributing to all men equal potentialities. The criterion of equality was not to be their religious or national heritage but their *Bildung*. Theirs was a cosmopolitan vision of mankind—evoked by the writings of the Prophets—in which national and religious differences would be transcended. *Bildung*, which nurtured the enfolding of the individual personality, was the mechanism by which this would be accomplished. *Bildung* was a particularly attractive notion for German Jews because they saw it as a means to surmount the gulf between their own history and the German tradition and to be able "to stress what united rather then what divided."[22] The classical concept of *Bildung* largely determined the postemancipatory Jewish identity. It seemed ready-made for their needs: everybody could attain *Bildung* through education and self-development. *Bildung* was to be the means by which German Jews were to integrate themselves into German society.

For the German Christians of the new *bürgerliche* class that was vying for power, the concept of *Bildung* fulfilled two vital functions: it helped legitimize them, by differentiating them from the upper and lower classes, and it facilitated the creation of an elite that provided better civil servants for the Prussian state. But for emancipated German Jews, *Bildung* came to serve a different function. It eventually became detached from the individual and became a kind of religion: "the worship of the true, the good and the beautiful."[23]

By midcentury, the belief in individualism and the potential of human reason—the central tenets of *Bildung*—became eroded as nationalism, with its myths and symbols, took its place and became the dominant ideology, not only in the popular mind but also among the mandarins. But emancipated German Jews resisted the narrowing of the vision. The more the older concept of *Bildung* came under attack, the more it became identified as the essence of emancipated, secularized German Jewry. *Bildung* became the faith of emancipated Jews, and their Jewish identity became defined by *Bildung*. This was particularly so after Bismarck's rise to power, with its attendant militarism, when the original concept of *Bildung* became transformed and became a rallying point for conservative opposition to university reform.[24] Mosse quotes Berthold Auerbach, considered by his fellow Jews to be one of the most representative Jews of the nineteenth century, as stating: "Formerly the religious spirit proceeded from revelation, the present starts from *Bildung*." The emancipated German Jews' secular religion of humanity had at its core *Bildung* and trumpeted the ideals of rationalism and progress. Auerbach characterized it as "an inner liberation and deliverance of man, his true rebirth; not through words or customs, but through his deeds, his character, the totality of his life, the cleansing and healing of all human labor."[25] Assimilation for these Jews meant a commitment to the Humboldtian notion of *Bildung*, a liberal outlook on society and a liberal position in politics. Many of them were attracted to the natural sciences and mathematics, in part because of the universalism of the two fields. Moreover, the norm of personal merit in these disciplines resonated with the individualism and cosmopolitanism of their Humboldtian *Bildung* tenets.

Throughout the nineteenth century, except for some rare exceptions, Jews were excluded from the civil service and the university professorate.[26] Although by the end of the century Jews nominally had the right to be appointed to university professorships, in 1909–10 fewer than 3 percent of the full professors at German universities were Jewish.[27] Access to university teaching careers still demanded conversion, particularly so in the humanistic faculties. Many Jews did convert[28] as indicated by the relatively large percentage of the professorate—around 9 percent—who had parents, grandparents, or great-grandparents who had converted.

Physiology and the allied medical sciences had attracted a number of such people, for example Eduard Hitzig, Julius Sachs, and J. R. Ewald.[29] The reason for this is not hard to find. Medical reforms instituted after the Revolution of 1848 had granted physicians a legal monopoly over medical care and had required the members of the profession to be trained at universities. Their accreditation was contingent on satisfactorily meeting all the demands of a curriculum determined and administered by the professorate. Midcentury also coincided with a transformation of German physiology. In a famous manifesto, Hermann Helmholtz, Emil du Bois-Reymond, Ernst

Brücke, and Carl Ludwig had vowed to reconstitute physiology on a physico-chemical foundation. The success of their reductionist and deterministic approach gave physiology a status of equal rank with physics. As a result of their productive program of research, the field became an autonomous scientific discipline and a core element in the new curriculum for the education of German physicians. Rudolf Virchow, in particular, had forcefully argued the value of physiology in securing the scientific foundations of the new medicine. He also argued that exposure to the scientific methodology required for understanding experiments in physiology would provide valuable mental training for the physician in his medical practice. In fact, starting in the late 1850s, physiology replaced logic in the Prussian medical examination.[30] Between 1853 and 1873, professorships of physiology were established at almost all the German universities, and these provided new niches for academically inclined converted Jews. And even though they now were civil servants, many of them retained their Enlightenment-inspired, liberal outlook.

Toward the end of the century, as the German mandarinate became more conservative, there remained a small subset that adhered to the original Humboldtian vision. Its members were principally mathematicians, physical scientists, and medical scientists, the latter often Jews or descendants of converted Jews. Albrecht Bethe, Hans's father, was trained as a zoologist and physiologist. As he had to obtain an M.D. in order to teach physiology he came into contact with this circle of medical scientists, many of whom were emancipated Jews, or ones who had converted. There was a resonance between his moral, social, and liberal political views and those of these emancipated or converted German Jews, the *Kulturträger* of enlightened, cosmopolitan *Bildung*. He eventually married a woman of Jewish descent. Hans would similarly find attractive the warmth, liberality, and cosmopolitanism of the Ewald household. Ella Ewald, the wife of Paul Ewald and the mother of Rose, whom Hans married, came from a long line of liberal, emancipated Jews.[31]

It is interesting to note that mathematics constituted the only discipline in which Jews were appointed to ordinary professorships without having to convert. Thus the mathematician Carl Gustav Jacobi was the first Jew to hold a full professorship in Prussia after the emancipation of the Jews there in 1812;[32] his contemporary, Moritz Abraham Stern, was appointed *ausserordentlicher Professor* in Göttingen in 1849, and ordinary professor in 1859. By the end of the 1860s, many more Jewish mathematicians held full professorships: Lazarus Fuchs, Leo Königsberger, Leopold Kronecker (who was made a member of the Berlin Academy of Sciences and could therefore teach at the University of Berlin by virtue of this), Paul Gordan, Max Noether, Adolf Hurwitz. The fact that it is much easier to recognize outstanding ability and talent in mathematics than in the humanities made it easier to argue for their appointment. But it required courage for their colleagues to recommend them. In the last quarter of the century, the mathematician Felix Klein—who was not a Jew—took on Humboldt's mantle and was responsible for the appointment of several outstanding Jewish mathematicians at German universities.

Klein had noted the onslaught of specialization and of industrialization and had come to the conclusion that philology and classical studies could no longer provide the coherent and harmonious foundations for the curriculum of pre-university studies that they had in Humboldt's time. He proposed that pure mathematics replace classical learning as the core of the secondary school curriculum because it could create a unified picture of nature by linking different areas of the sciences.[33]

Klein was a visionary. He altered the social relations in mathematics and melded German mathematicians into a community of interacting practitioners. After his appointment in Göttingen, he bound the Göttingen scientific community by mathematical ties. He gave the Göttingen science and applied science enterprise a sense of unity by virtue of mathematical tools. His commitment to Humboldtian ideals was explicit: exceptional ability and excellence were the only requirements for an appointment to the Göttingen mathematics faculty.[34] It did not matter whether the person was Jewish, male, or female. And, in fact, Klein did recommend Adolf Hurwitz, Hermann Minkowski, Paul Gordan, and Emmy Noether for appointments to the Göttingen faculty; but Friedrich Althoff, the education minister who had to approve all appointments, and the Göttingen faculty objected, and only Minkowski was appointed.

There is another thread that tied Felix Klein to Wilhelm von Humboldt: his belief in a preestablished harmony. With Klein and his fellow mathematicians, the Leibnizian preestablished harmony became more specific. It became a preestablished harmony between physics and mathematics and the foundation of their pantheistic faith. Klein and David Hilbert paid lip service to it, and Minkowski ended his famous paper on the four-dimensional representation of special relativistic space-time with a rhapsodic paean to preestablished harmony. The great interest by physicists in variational principles around the turn of the century (e.g., by Hermann von Helmholtz, Max Planck, Martin Schwarzschild, Paul Ehrenfest, Albert Einstein) is another manifestation of a belief in this newer version of preestablished harmony.

One of Felix Klein's closest associates during the 1890s was Arnold Sommerfeld. He too came to believe in a preestablished harmony between physics and mathematics. He passed this belief on to his gifted students, among them Wolfgang Pauli and Hans Bethe.

BECOMING A PHYSICIST: ARNOLD SOMMERFELD

Bethe's talents, particularly his numerical and mathematical abilities, had manifested themselves early. By the time Hans finished the Gymnasium, he knew he wanted to be a scientist, and his poor manual dexterity steered him first into mathematics and then into theoretical physics. After completing two years of studies at the University of Frankfurt, he went to Arnold Sommerfeld's seminar in Munich. Bethe obtained his doctorate in 1928 *summa cum laude*. After a brief stay in Stuttgart as Paul Ewald's assistant, he returned to Munich to do his *Habilitation* (begin his professional academic career) with Sommerfeld. During the academic year 1930/31 he was a Rockefeller fellow at the Cavendish Laboratory in Cambridge and in Rome at Enrico Fermi's institute. In 1932 he again spent six months in Rome working with Fermi. By 1933 he was recognized as one of the outstanding theorists of his generation. His book length *Handbuch der Physik* articles on the quantum theory of one- and two-electron systems and on the quantum theory of solids became classics as soon as they were published.[35] In April 1933, after Hitler's accession to power, Bethe was removed from his position in Tübingen because he had two Jewish grandparents. He went to England, and in the fall of 1934 he accepted a position at Cornell University. He arrived there in February 1935 and has been there ever since. In 1937, Rabi wrote the following letter to F. Wheeler Loomis, the chair of the Department of Physics at the University of Illinois, who was trying to get Bethe to come there:

> To write about the qualifications of H. A. Bethe is a very pleasant occupation although it does tend to somewhat exhaust one's supply of superlative adjectives. . . . Sommerfeld regarded him as his best student after Heisenberg and Pauli. . . . His coming to Cornell revitalized that physics department as one can easily see from publications in the *Physical Review*.
>
> Bethe possesses a thorough command of all the tools of theoretical physics and an amazing virtuosity in their application. He has contributed to all branches of modern physics from the solid state to nuclear physics. He writes with great facility and effectiveness as shown by his two long articles which are really

large sized books in the *Handbuch der Physik* and by the articles in the *Reviews of Modern Physics.*

His English is excellent. In my opinion he is just about the clearest lecturer I have heard. He possesses this quality on every level from the most elementary treatments to the most complicated. He is very adaptable and has become quite Americanized in the short time he has been in this country.

He has a great interest in experimental physics and understands experimental problems thoroughly. His help in this connection has been very valuable to the Cornell physicists.

All in all I should place him among the five best physicists in the United States. My opinion of him can best be summed up by saying that I wished we had him at Columbia.[36]

During World War II, Bethe worked on armor penetration, radar, and helped design atomic weaponry. He was a member of the Radiation Laboratory at MIT from 1942 until the spring of 1943, when he joined Oppenheimer at Los Alamos and became the head of the Theoretical Division. After the war he became deeply involved in the peaceful applications of nuclear power, in investigating the feasibility of developing fusion bombs and ballistic missiles, and in helping to design them. He served on numerous advisory committees to the government, including the President's Science Advisory Committee (PSAC). In 1968 he won the Nobel Prize for his theoretical investigations in 1938 explaining the mechanism of energy production in stars. His most recent researches have been concerned with the life cycle of supernovas and the properties of the neutrinos involved in the fusion processes in the sun. He has been and continues to be an enormously productive scientist.

The community of physicists recognizes two kinds of genius. There are those who are deemed geniuses by the originality of their thought, who are placed altogether above and beyond the established norms by the startling novelty of the style of their thought, who, like magicians, seem to pull rabbits out of their hats. Feynman was regarded as that kind of "genius."[37] Then there are those, like Bethe, who are considered off-scale because of the magnitude of their contribution to developing a body of knowledge, *whose ability to synthesize overwhelms.* Sometimes off-scale people dare too much and run into their natural limits—perhaps this was the case with Oppenheimer in the late 1930s.[38] Others, like Bethe, do not encounter these limits because their vision is more modest and limited. They are content to create novelty from existing structures and resources. They remain creative to the end of their lives.

Bethe is a verbal thinker rather than a visual one. His modes of thought are logical, incremental, additive. Indeed, he makes great use of the vast store of information accurately imprinted in his prodigious memory, and invariably, when tackling a problem, relies on the numerous models of analogous systems or processes that he has previously analyzed and solved in order to fashion an applicable approach or model. The stored "pieces" yield clues and insights and are reassembled in new ways. Often, the reconnected parts yield a synthesis that result in new coherence, and with it novelty and fructification. Bethe has not invented new modes of expression—he is perfectly content to learn and use the extant modalities. For him, physical theories are simplified, algorithmic representations of the physical world, and mathematics is the language used for their symbolic representations. These representations are constrained by the physical structure of the world. And Bethe delights in the constraints and in the inner logic of mathematics that allows new information and new ways of understanding and looking at the world to emerge from the manipulations of its symbols.

The constraints of nature are severe. Nature allows the physicist to carve out and abstract domains that are populated by seemingly ahistoric entities whose dynamics are accurately represented by time translation invariant laws—that is, laws that are seemingly immutable on enormous timescales. But there is a great deal of leeway in how these laws are expressed at any given time. It is in this freedom and from the analysis of the choices that are made that we can recover "the author-in-the-works" for the case of the scientist. One of the striking features of Bethe's corpus of scientific works is what in literary criticism would be called his sustained development of a given plot. This is in contrast to Oppenheimer's prewar corpus, which was perhaps more brilliant but also more episodic and fractured.[39] Bethe's *Handbuch* articles were triumphant *grands recits* of the developments of nonrelativistic quantum mechanics. They not only revealed Bethe's mastery of the subject matter—often more thorough, precise, and insightful than the authors of the quoted articles—they also reflected facets of his character: conscientiousness, reliability, hard work, and thoroughness. This has remained true of all his review articles. All his scientific productions exhibit these qualities as well. Bethe has enormous *Sitzfleisch*—perseverance, resolve, stick-to-itiveness—and huge powers of sustained concentration that he brings to bear in his explorations of any given subject matter and any problem he tackles.

In Bethe's scientific researches, there is at work a consciousness that can be discerned in the way he explores, orders, and transposes a set of basic themes. One of these is the identification of the "elementary" entities—the building blocks—of the various realms of nature that he has investigated and the characterization of the interactions between them in as "fundamental" a way as possible. This of course has been the research program of physicists since the beginning of the century. But Bethe is interested in more: he wants to explain how new stable structures—new wholes—can be built out of the elementary entities. He has done so for atoms, molecules, and solids, for nuclei, for stars. Initially, it is the explanation of the stability of the composite structures that interests him—but eventually it is their history, i.e., the evolution of such systems that concerns him. The most obvious examples are his researches on stellar evolution and more recently his investigations of the death throes of supernovas.

An analysis of Bethe's work on stellar energy in the late 1930s reveals another one of his strengths: his ability to live with uncertainty and partial knowledge. His explanation of the mechanism for energy generation in stars grew out of his participation in April 1938 at the third Washington conference on theoretical physics. These conferences had become annual events that were organized by George Gamow, Edward Teller—at the time, one of Bethe's closest friends—and Merle Tuve. He initially had refused the invitation to participate in a conference devoted to stellar energy and stellar structure because these were areas outside his immediate research interests, but persistence on the part of Teller finally convinced him to attend. Following the meeting, Subrahmanyan Chandrasekhar, George Gamow, and Merle Tuve issued a brief report in which they indicated that "some interesting conclusions had been reached" at the conference concerning nuclear transformations as the source of energy production in stars.[40] They indicated that it was generally agreed that the so-called *Aufbauhypothese*—the suggestion that the production of the heavier elements from hydrogen is continually taking place in the interiors of stars and that in the process energy is liberated—had run into difficulty. Chandrasekhar, Gamow, and Tuve then suggested another reaction—a triple collision between an alpha particle and two protons—as a possible mechanism for nucleosynthesis and generating energy. They also reported that the reaction

$$^1H + {}^1H \rightarrow {}^2H + \beta^+ + \text{neutrino}$$

first suggested by Carl von Weizsäcker as "another possibility" for energy generation and the production of deuterium, and whose rate of reaction had been calculated by Bethe and Charles Critchfield, had been discussed at the conference. "It seems that the rate of such a reaction under the conditions in stellar interiors would be enough to account for the radiation of the sun, though for stars much brighter than the sun other more effective sources of energy are required."

Their note in *Nature* indicates that in the minds of most of the participants, the problem of nucleosynthesis was conflated with the problem of energy generation. Bethe, on the other hand, went back to Cornell and *separated* the two problems. He advanced two sets of reactions—the proton-proton and the carbon cycle—that were to account for energy production in stars like the Sun. The second depended on the presence of carbon in the star. At that time there was no way to account for the abundance of carbon in stars, that is, it was not at all clear what nuclear reactions in stars between elements lighter than carbon could produce this element. However, the presence of carbon in stars had been corroborated by their spectral lines in stellar atmospheres. Bethe accepted this fact and proceeded to compute the characteristics of stars nourished by the two cycles and found that the carbon-nitrogen cycle gives about the correct energy production in the sun.[41]

The problems of nucleosynthesis and that of energy production in stars would later once again be brought together. In 1938 the separation of the two problems was an important step for the resolution of both.

WHOLENESS AND STABILITY

The explanation of the emergence of new structures from the synthesis of their components, and the explanation of their stability, are recurring foci in Bethe's works. The search for wholeness and stability and the expression of integrity have also been constant themes in his personal life. In fact, one can try to comprehend Bethe's life as a search for integration and integrity—for stability and unity under the conditions of growth and evolution. Bethe grew up in a world that had been sundered and was unstable. World War I and Weimar Germany shaped the political and economic context of his teens. His personal world was similarly in disequilibrium. As a child he was often sick and had to be privately tutored. During World War I, in the summer of 1916, when he was ten years old, he contracted tuberculosis. That fall he was sent to a children's home in Kreutznach, where he stayed until February 1918.[42] He essentially had no close friends until he entered the Gymnasium in Frankfurt. His parents did not entertain much and had few visitors; he characterized their home as a "quiet place; a very quiet place." Nor did they have any connection with a religious community that might have given some ballast to their social world. Neither Hans's father nor his mother were religious, and her baptism is an indication of her distaste for the Jewish religion.[43] As Hans grew up, his parents drifted apart. Periodically, his mother needed to go to a "rest home"; eventually, his parents divorced in 1927. Thereafter, Hans took care of his mother and looked after her needs. During the 1930s he regularly visited her for extended periods during the summer. He got her out of Nazi Germany in 1939, after *Kristallnacht* had finally convinced her that she was no longer safe there. After she arrived in the United States he supported her stay in a convalescent home in Queens (New York) and later in Ithaca.

Hans was close to his father, and it was his father who introduced him to the world of science and encouraged him to pursue his interests in mathematics and the sciences. It was also through his father that Hans was introduced to what Alfred North Whitehead has called the "fallacy of misplaced concreteness." His father's physiological researches involved precisely "the taking of an object out of its natural setting, out of the normally rich context of mediations that reality offers, in order to consider its intrinsic properties in isolation of the world around it."[44] Coupled to this Cartesian outlook was also a belief that the world could be recomposed marginally and additively. Martin Krieger has insightfully noted that the calculus is the emblematic paragon of that worldview.[45] This conceptualization of the world shaped modern science. It spilled over into the models of the social world and polarized the ideology of classical liberalism that depicted even the political rights of individuals in isolation from society as a whole. In ethics, in the German context, the Kantian categorical imperative served as the guiding principle for moral behavior. Here, too, the emphasis was on individual action. Bethe also obtained from his father an introduction to neo-Kantianism and its implications. His father's metaphysics, like that of most of the German professorate in the medical sciences at the time, was predicated on the premise that beyond the inherently limited range of man's perceptions lay a world of objects that was ultimately unknowable. Rationality could provide some measure of control; it could even provide some understanding, but ultimate knowledge of the nature of things was beyond man's reach. Since the overall relation between phenomena was believed to be ultimately unknowable, scholars should confine themselves to carefully establishing and clarifying the data appropriate to their given discipline. Every field of knowledge was a kind of calculus, a rigorous rational system that was referable to a set of entities—the ontology of that more or less self-contained science. This outlook fostered the academic specialization that characterized German universities during the latter half of the nineteenth century. This atomization also required them to confront their dedication to the ideal of *Bildung*. Despite the centrifugal pull of specialization, the German professorate renewed its commitment to *Bildung*, partly for social reasons stemming from its changing class position. They argued that only the professorate—by virtue of *Bildung*—could fashion a coherent and unitary *Weltanschauung* that would meld into a solidary whole the nation's warring factions riven by political and economic differences due to the rapid industrialization and urbanization that Germany was undergoing.[46]

The young Hans responded to the intellectual and social world around him by adopting a Weberian stance: he would deal with the world rationally, to the utmost limits that rationality would allow. Already as a teenager he honed his ability to ascertain the causal relationships that operated in any situation that involved well-defined ends. He also willed to limit personal pleasure in order to increase efficiency in securing ends. This rational approach worked well in the formal, narrowly circumscribed milieu of German academia that constituted his professional world until Hitler ousted him from it.

If self-containment and self-reliance mark Bethe's initial stance, his first insights into the meaning and the comforts of community came from his interactions with the circle of friends he acquired while in Gymnasium and his stay at the University of Frankfurt. Most of these friends came from emancipated Jewish families for whom the tenets of the Enlightenment were their secular religion. Their outlook was cosmopolitan, liberal, and they were open to the possibilities of the world. In this too he was following the footsteps of his father, who had married a woman with a similar cultural background. The full force of the warmth and the strength of that community struck Bethe when in 1929 he became Paul Ewald's assistant in Stuttgart. Ewald, whose wife was the niece of a famous and influential reform rabbi, opened his home to the young Bethe and he became a frequent visitor. Ten years later he married Rose, one of Ella and Paul Ewald's daughters.

Through Sommerfeld, whose seminar in Munich he entered in 1926, Bethe was introduced to the Göttingen view of the preestablished harmony between physics and mathematics. The unification of geometry and gravitation that Einstein had molded in his general theory of relativity had deeply impressed Sommerfeld. But more important for the young Bethe was the unity and integrity that Sommerfeld embodied. Sommerfeld's powers were such that he had a command of all of physics and most of mathematics until the early 1930s.

The mastery of the various branches of physics was Sommerfeld's anchor in integrity.

Sommerfeld was a forceful and charismatic figure, and though very much the "Herr Geheimrat," nonetheless the atmosphere in the seminar was characterized by the intellectual give-and-take between him and his students and assistants. Among them are to be found many of the outstanding theorists of their generation: Peter Debye, Paul Epstein, Paul Ewald, Max von Laue, Wolfgang Pauli, Werner Heisenberg, Gregor Wentzel, Fritz London, Bethe.[47] It is a measure of the man that he never felt threatened by the brilliant students he trained. Sommerfeld learned from his *Doktoranden* and assistants and collaborated with them. After they left the seminar they wrote to him and kept him informed of the latest developments. His students fashioned new tools that he, the master craftsman, learned to use. He thus continued to grow and to adapt to the new topography of theoretical physics. After the advent of quantum mechanics, he was one of the first to integrate the new materials into a textbook.[48]

It was in Munich that Bethe anchored his self-confidence. He discovered there his exceptional talents and his extraordinary proficiency in physics. Sommerfeld gave him indications that he was among the very best students who had studied with him. This self-confidence in matters of physics quickly extended to other matters.[49] This self-confidence also gave him the courage to face the world in moral terms.

Sommerfeld had a deep commitment to the school of theoretical physics he established in Munich. He had several opportunities to leave Munich for more prestigious positions,[50] yet he remained there. Bethe learned from Sommerfeld what commitment to an institution and to a tradition meant. Bethe would emulate his teacher's comportment, and at Cornell he too built a school of physics where he trained and influenced some of the outstanding theoretical physicists of their generation: Emil Konopinski, Morris Rose, Robert Marshak, Richard Feynman, Freeman Dyson, Richard Dalitz, Edwin Salpeter, Geoffrey Goldstone, Robert Brout, David Thouless, Peter Carruthers, Roman Jackiw, John Negele.[51] After World War II, Bethe received many lucrative professorial offers. In particular, Rabi tried very hard to have him come to Columbia. He turned down all these invitations. One offer was particularly meaningful: a request that he succeed Sommerfeld in Munich. Bethe's letter to Sommerfeld declining the offer is a moving testimony to his growth and maturation:

> I am very gratified and very honored that you have thought of me as your successor. If everything since 1933 could be undone,

I would be very happy to accept this offer. It would be lovely to return to the place where I learned physics from you, and learned to solve problems carefully. And where subsequently as your *Assistent* and as *Privatdozent* I had perhaps the most fruitful period of my life as a scientist. It would be wonderful to try to continue your work and to teach the Munich students in the same way as you have always done: With you one was certain to always hear of the latest developments in physics, and simultaneously learn mathematical exactness, which so many theoretical physicists neglect today.

Unfortunately it is not possible to erase the last 14 years. My father-in-law has written to you about that already—and I believe he expressed my feelings very well. For us who were expelled from our positions in Germany, it is not possible to forget. The students of 1933 did not want to hear theoretical physics from me (and it was a large group of students, perhaps even a majority), and even if the students of 1947 think otherwise, I cannot trust them. What I hear about the nationalistic orientation of students at many universities starting up again, and about many other Germans as well, is not encouraging.

Perhaps still more important than my negative memories of Germany, is my positive attitude toward America. It seems to me (already for many years) that I am much more at home in America than I ever was in Germany. As if I was born in Germany only by mistake, and only came to my true homeland at age 28. The Americans (nearly all of them) are friendly, not stiff or reserved, nor brusque (*gar ablehnend*), as most Germans. It is natural here to approach all other people in a friendly way. Professors and students relate in a collegiate way without any artificially erected barrier. Scientific research is mostly cooperative, and one does not see competitive jealousy between researchers anywhere. Politically most professors and students are liberal and reflect about the world outside—that was a revelation to me, because in Germany it was customary to be reactionary (long before the Nazis) and to parrot the slogans of the German National (*Deutschnationaler*) party. In brief, I find it far more congenial to live with Americans than with my German *Volksgenossen*.[52]

On top of that, America has treated me very well. I came here under circumstances which did not permit me to be very choosy. In a very short time I had a full professorship, probably more quickly than I would have gotten it in Germany if Hitler had not come. Although a fairly recent immigrant, I was allowed to participate in work and to have a prominent position in military laboratories. Now, after the war, Cornell has built a large new nuclear physics laboratory essentially "around me." And 2 or 3 of the best American universities have made me tempting offers.

I hardly need mention the material side, insofar as my own salary is concerned and also the equipment for the Institute. And I hope, dear Sommerfeld, that you will understand: Understand what I love in America and that I owe America much gratitude (disregarding the fact that I like it here). Understand, what shadows lie between myself and Germany. And most of all understand, that in spite of my "no" I am very grateful to you for thinking of me.[53]

In 1953 Oppenheimer invited him to join the Institute for Advanced Study. In declining the invitation, Bethe wrote him:

Dear Robert,

. . .

I want to thank you once more for your offer, for the way in which you made it to me, and for everything you have said and written to me on this matter after the first offer. Since you first talked to me, you have continued to make work at the Institute still more attractive to me, so that my decision has indeed been a very hard one.

Nevertheless, after more than eighteen years, I feel very much a part of Cornell University. Much of the Physics Department as it now exists has developed under my influence, and most of the members of the staff we now have have been brought together by the work of myself and some close friends. When I first mentioned the possibility of leaving, everybody impressed on me the extent of the changes in Physics here if I were to leave. I had anticipated some efforts to hold me here but it far exceeded my expectations.[54]

In addition to Sommerfeld, the other great formative influences on Bethe were Enrico Fermi and, to a lesser extent Patrick Blackett.

During 1930, Bethe spent a few months in Cambridge and a few months in Rome on a Rockefeller fellowship. The Cambridge experience was exhilarating. Bethe found the openness and cordiality of his British hosts, especially that of Ralph Fowler and Blackett, most engaging and attractive. Blackett was the first experimentalist with whom Bethe interacted strongly, and the interaction proved formative and seminal. Bethe's physics thereafter always reflected a close association with experimentalists and experimental work. Blackett, at the time, was particularly interested in Bethe's work on the energy loss of charged particles in their passage through matter, because Bethe's calculations could be used to identify cosmic particles by the tracks they left in Blackett's cloud chambers. And Bethe responded by making detailed calculations that proved very helpful to Blackett.

Evidently, the Cambridge surroundings allowed the rigidity that a German education bestowed on scholars to be shed, for the year 1931 opened with the appearance of a startling *kurze Originalmitteilung* in *Naturwissenschaften* entitled "Concerning the Quantum Theory of the Absolute Zero of Temperature." The note was signed by G. Beck, H. Bethe, and W. Riezler, three postdoctoral fellows at the Cavendish Lab. Coming on the heels of Eddington's attempt to explain the numerical value of the fine structure constant, the note read as follows:

> Let us consider a hexagonal lattice. The absolute zero of the lattice is characterized by the fact that all degrees of freedom of the system are frozen out, i.e. all inner movements of the lattice have ceased, with the exception, of course, of the motion of an electron in its Bohr orbit. According to Eddington every electron has $1/\alpha$ degrees of freedom where α is the fine structure constant of Sommerfeld. Besides electrons our crystal contains only protons and for these the number of degrees of freedom is obviously the same since, according to Dirac, a proton is considered to be a hole in a gas of electrons. Therefore to get to the absolute zero we have to remove from the substance per neutron (= 1 electron plus 1 proton; our crystal is to carry no net charge) $2/\alpha - 1$ degrees of freedom since one degree of freedom has to remain for the orbital motion. We thus obtain for the zero point temperature $T_0 = -(2/\alpha - 1)$ degrees. Putting $T_0 = -273°$, we obtain for $1/\alpha$ the value 137 in perfect agreement within the limits of accuracy with the value obtained by totally independent methods. It can be seen very easily that our result is independent of the particular crystal lattice chosen.

Since in those days papers in respected scientific journals were read with absolute trust in the honorable intentions of the authors and the editors, it took a while for the community to realize that the *Naturwissenschaften* had been fooled and that the paper was a prank. Arnold Berliner, the editor of the *Naturwissenschaften*, was not amused. Nor was Sommerfeld. Berliner demanded an apology and on March 6 there appeared a "Correction" in *Naturwissenschaften*:

> The note by G. Beck, H. Bethe and W. Riezler, published in the January 9 issue of this journal, was not meant to be taken seriously. It was intended to characterize a certain class of papers in theoretical physics of recent years which are purely speculative and based on spurious numerical agreements. In a letter received by the editors from these gentlemen they express regret that the formulation they gave to the idea was suited to produce misunderstanding.[55]

The episode attests to Bethe's sense of humor, whose expression had been restrained by the stolidity and formality of the German setting.

After his stay in Cambridge in 1930, Bethe went to Fermi's institute in Rome. Fermi helped Bethe free himself from the rigorous and exhaustive approach that was the hallmark of Sommerfeld. From Fermi, Bethe learned to reason qualitatively, to obtain insights from back-of-envelope calculations, to think of physics as easy and fun, as challenging problems to be solved. In Rome he discovered what "lightness" meant. He there was also exposed to a freer and more informal mode of interaction between a professor and his students than had been the case in Munich.

Bethe's craftsmanship is an amalgam of what he learned from these two great physicists and teachers, combining the best of both: the thoroughness and rigor of Sommerfeld with the clarity and simplicity of Fermi. This craftsmanship is displayed in full force in the many "reviews" that Bethe has written. His first "review" was the result of Sommerfeld asking him to collaborate in the writing of his *Handbuch der Physik* entry on solid state physics. This was the first of many subsequent reviews, of which the two *Handbuch* entries[56] and the Bethe *Bible* on nuclear physics are merely the most famous.[57] I call them reviews, but this is a misnomer. They are really syntheses of the field under review, giving them coherence and unity, charting the paths to be taken in addressing new problems. They usually contain much that is new, materials that Bethe worked out in the preparation of the essay.

Before World War II, Bethe derived a sense of wholeness from his mastery of all of physics, in particular the newer fields of physics that were being probed by the new quantum mechanics and by the new apparatus and instrumentation in the nuclear realm, such as cloud chambers, cyclotrons, and beta-ray spectrographs. For the most part, quantum mechanics could successfully account for the structures that were observed in the microscopic realm of nature: nuclei, atoms, molecules, and how these aggregated into macroscopic structures such as solids and stars. Bethe contributed importantly to each of these fields, often shaping newly emerging subdisciplines. But it should be stressed that he never attempted to "unify" these fields in the sense of trying to find a unitary theory that would encompass all the phenomena in all these realms.

Bethe is the supreme example why theoretical physicists proved to be so valuable in the war effort. He is as much at home in applied mathematics as he is in the many domains of theoretical physics that he has mastered. No calculation fazes him, and if analytic solutions cannot be obtained he will resort to doing extensive numerical work. Because he understands every facet of the experimental practice, he can interpret as well as suggest experiments. But it is his ability to translate his understanding of the microscopic world—that is, the world of nuclei, atoms, molecules—into an understanding of the macroscopic properties and behavior of materials, and into the design of macroscopic devices, that rendered his services so valuable at Los Alamos and later on to industry. Bethe's mastery of quantum mechanics and statistical mechanics allowed him to infer the properties of materials at the extreme temperatures and pressures that would exist in an atomic bomb. Bethe, Fermi, and the other physicists on the Manhattan Project converted their knowledge of the interaction of neutrons with nuclei into diffusion equations, and the solutions of the latter into reactors and bombs.

Bethe left Los Alamos a generalist in applied science while retaining his command of "pure" physics. Thereafter, his unique competence in all these fields made him an invaluable resource to industry and government laboratories. He frequently returned to Los Alamos as a consultant and most years spent a fraction of the summer there. Often his technical skills as a generalist were put at the service of friends. Bethe sees the spectrum of human abilities hierarchically ordered. He has a very strong sense of his own abilities and capabilities and almost unconsciously assesses and gauges those of the persons he comes in contact with. He can therefore tell in what ways he can be helpful to them. Friendships often determined the projects he worked on at Los Alamos. The problems had to be interesting—but there were a multitude of those! His joining a project often was a tacit expression of his affection and respect for the individuals involved and an unspoken statement of his solidarity with their community.[58]

After World War II, the rapid expansion of physics, the concomitant explosion of knowledge, and Bethe's extensive consulting activities all combined to make it impossible for him to keep abreast of all developments in physics and to retain his mastery of all of physics. One of the turning points in his life occurred in the mid-1950s, when he decided not to work in high energy physics anymore, but to concentrate his researches in nuclear physics. To maintain his integrity, Bethe must master the fields he works in. It is inconceivable for him to be a dilettante; it is impossible for him to do superficial work. This contraction in the intellectual sphere coincided with painful events at home. Bethe confronted these realities and in the process grew.

One can identify three fairly well delineated "stages" in Bethe's life until the mid-fifties. In the period from 1906 until 1933, it was German culture and German institutions that molded him. The two *Handbuch der Physik* articles are the fruition of stage 1. From 1934 until 1940, Cornell was his haven. The Bethe *Bible* and his solution of the problem of energy generation in stars epitomize the capacities of the mature scientist, of the scientist who has helped shape the new field of subatomic phenomena. The third period, which begins with the outbreak of World War II, sees him acquiring new powers at the Radiation Laboratory at MIT and at Los Alamos. The postwar years from 1946 to 1955 constitute one of the most exhilarating phases of his life, both scientifically and professionally. The stage of his activities has become national and international. He is at the center of important new developments in quantum electrodynamics and meson theory. He helps Cornell become one of the outstanding universities in the world. He is a much sought-after and highly valued consultant to the private industries trying to develop atomic energy for peaceful purposes. He is deeply involved and exerts great influence in matters of national security. And he is happily married and the proud father of two very bright children. But the demands from his activities outside Cornell were enormous, the pace grueling, and they were exacting a heavy toll both at home and in his scientific researches. In 1955, Bethe went to Cambridge University to spend a sabbatical year there. It was a year of taking stock and of narrowing the focus of his scientific researches.

In 1958, in a review of Robert Jungk's *Brighter than a Thousand Suns*, Bethe revealed some of his despondency.[59] He there recalled the "golden times" of the 1920s and 1930s when science for him was "a great spiritual adventure." All experimental discoveries were made with small, rather inexpensive apparati, every detail of which could be understood. One could then know all of physics. And most importantly, "The physicists in all countries knew each other well and were friends. And the life at the centers of the development of quantum theory, Copenhagen and Göttingen, was idyllic and leisurely, in spite of the enormous amount of work accomplished." He confessed that he could not help but be nostalgic reading about these times in the book:

How it has all changed! There are now enormous accelerators, with large groups of scientists working on each, a wealth of detailed material is published in highly specialized journals every week so that it has become impossible to keep up with the literature even in the narrow part of nuclear physics. . . . The life of physicists has changed completely, even of those not involved in politics or in technological projects like atomic energy. The pace is hectic. Yet the progress of fundamental discovery is no faster, and perhaps slower, than in the thirties.

Between 1955 and the mid-1970s, Bethe's activities in "pure" science were concentrated in the study of nuclear matter. Although these researches were consistently of high quality, they cannot be characterized as outstanding. The pressure of his other activities—membership on PSAC and its subcommittees, consulting for Avco, GE, and other industrial firms—and crises within the home help explain the character of his scientific productions during this period. Since the mid-1970s, Bethe has been collaborating with G. E. Brown, and this association has resulted in renewed creativity and exceptional productivity. During the past decade, Bethe has contributed importantly to the elucidation of supernova explosions and to the solar neutrino problem. Brown is some twenty years younger than Bethe. Their collaboration started out as the joining of forces by two outstanding scientists whose talents complemented each other. Since then Bethe has become Brown's "postdoc," and he looks forward to becoming Brown's "graduate student."

As time went on, there has been a gradual shift in the position where Bethe places the fulcrum in his attempt to resolve the tension between individualism and community, between dissension and solidarity. In a world where no one can master all knowledge and skills, communities assume new dimensions as dynamic, generative enterprises that produce new knowledge, new tools, new generations of practitioners, and give coherence and meaning to the human enterprise.

For Bethe, the practice of science has been a spiritual quest, a search for understanding, human understanding. He believes that all understanding—and scientific understanding, in particular—requires that one test and risk one's convictions and prejudgments in and through encounters with others, and demands confronting what is new and alien. The scientific community provides the channels and the setting for such encounters. It is a community that relies on the virtues of honesty, tolerance, trust, truthfulness, and cooperation; one that guarantees tolerance for the views of others and the verification of their claims. It is the community that authenticates the integrity of our scientific understanding. Without a *moral* commitment on the part of scientists to be truthful and trustworthy members of this community, no important statements of fact could see the light of day and be stabilized.[60] Bethe has—consciously and conscientiously—endowed the two institutions he has been most closely associated with, Cornell University and its Newman Laboratory for Nuclear Studies, with these qualities.

These essential attributes of the scientific enterprise were a recurring theme in the public lectures Oppenheimer gave after the war. For Oppenheimer, science was "an area of collective effort in which there is a clear and well-defined community whose canons of taste and order simplify the life of the practitioner. It is a field in which the technique of experiment has given an almost perfect harmony to the balance of thought and action."[61] And to the question: What are the lessons that the spirit of science teaches us for our practical affairs? Oppenheimer gave the following answer:

> Basic to them all is that there may be no barriers to freedom of inquiry. Basic to them all is the ideal of openmindedness with regard to new knowledge, new experience and new truth. Science is not based on authority. It owes its acceptance and universality to an appeal to intelligible, communicable evidence that any man can evaluate.
>
> There is no place for dogmas in science. The scientist is free to ask any question, to doubt any assertion, to seek for any evidence, to correct any error. For scientists it is not only honorable to doubt; it is mandatory to do that when there appears to be evidence in support of the doubt.

Moreover, in characteristically American fashion, Oppenheimer wanted to draw lessons for the body politic from the operation of science:

> Our political life is predicated on openness. We do believe any group of men adequate enough or wise enough to operate without scrutiny or without criticism. . . . [T]he attitudes of mind, . . . the disciplines of spirit which grow naturally in the scientist's world . . . have grown there in part as a result of a humane and liberal tradition in political life and in part as a cause of that.[62]

I. I. Rabi, in an address at the opening exercises of Columbia University for the academic year 1950–51, had outlined the same vision and had stated what the tradition of science had taught him:

> It teaches us moderation and tolerance of ideas, not because of lack of faith in one's belief, but because every view is subject

Figure 1. The young J. Robert Oppenheimer. (Courtesy AIP Niels Bohr Library, *Physics Today* Collection)

Figure 2. Dirac, Suguira, and Oppenheimer, Göttingen, 1928. (Courtesy AIP Niels Bohr Library, Uhlenbeck Collection)

Figure 3. Oppenheimer (second from left in back) at a weekly colloquium in Los Alamos. (Los Alamos Scientific Laboratory. Courtesy AIP Niels Bohr Library)

Figure 4. Oppenheimer in the postwar period. (*Bulletin of Atomic Scientists*. Courtesy AIP Emilio Segrè Visual Archives)

Figure 5. Oppenheimer in the early 1960s. (Courtesy AIP Emilio Segrè Visual Archives, *Physics Today* Collection)

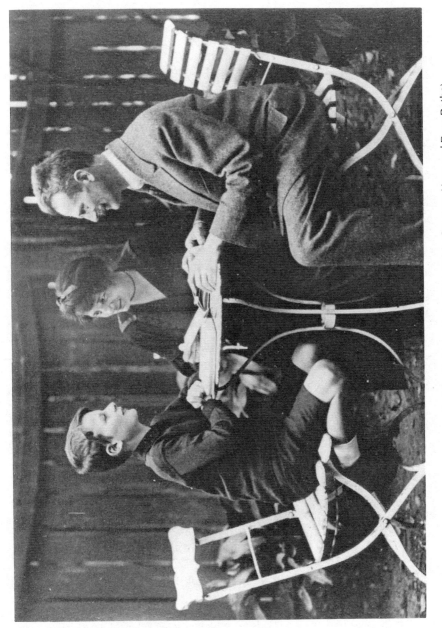

Figure 6. The young Hans Bethe with his parents. circa 1920. (Courtesy Hans and Rose Bethe)

Figure 7. Bethe (front center) at the 1939 Washington Conference on Theoretical Physics. Others in the photo include Niels Bohr, Eugene Wigner, John Wheeler, I. I. Rabi, Edward Teller, and George Gamow. (Courtesy Department of Manuscripts and University Archives, Cornell University Libraries, Ithaca, New York)

Figure 8. Bethe at Los Alamos during World War II. (Courtesy AIP Emilio Segrè Visual Archives)

Figure 9. Bethe giving a colloquium, circa 1960. (Courtesy Department of Manuscripts and University Archives, Cornell University Libraries, Ithaca, New York)

Figure 10. Bethe on the day he learned that he was awarded the 1968 Nobel Prize in physics. (Courtesy Department of Manuscripts and University Archives, Cornell University Libraries, Ithaca, New York)

to change and every truth we know is only partial. The strange thought or custom may still be valid.

It teaches cooperation not only among people of the same kind, but also of the most diverse origins and cultures. Science is the most successful cooperative effort in the history of mankind.

Science inspires us with a feeling of hopefulness and of infinite possibility. . . . Science shows us that it is possible to foresee and to plan and that we can take the future into our own hands if we rid ourselves of prejudice and superstition. . . .

Science teaches us self-discipline. . . .

These lessons can be multiplied to cover almost the entire range of human activity, because science is itself a contemporaneous living thing made by men for man's edification and entertainment.

Science is great fun. . . . Science is a great game. It is inspiring and refreshing. The playing field is the universe itself. The stakes are high, because you must put down all your preconceived ideas and habits of thought. The rewards are great because you find a home in the world, a home you have made for yourself.[63]

Bethe would have made similar assertions and certainly shared Rabi's sentiments.

As he has grown older, Bethe has valued ever more his membership in the various communities that are committed to these ideals and that have allowed him to be productive, to continue growing, and to keep on constantly becoming.

Bethe's continued craftsmanship and mastery of physics have been his anchor in honesty and integrity; and integrity has resulted in an integrated personality. It is one of Bethe's striking characteristics that there is only one of him—in contrast to Oppenheimer. Bethe is the same whether dealing with a student, with a colleague, with the president of Cornell, or with a senator in Washington. It is an expression of the integrated nature of his self; it is an expression of his integrity. And this despite the fact that of course no person is whole; no person is not in some ways fractured; no person is not the reconstituted entity of the fragmented pieces of his or her past.

Bethe's striving for wholeness has been constant throughout his life. In the Göttingen tradition that his teacher Sommerfeld embodied, it was believed that this goal could be achieved through intellectual commitments; that the unity one seeks in the laws of nature would extend to the human realm—in particular, that this sense of unity would be achieved when individuals and the communities of which they are members all have integrity; that their interactions would enhance their stability and well-being.

Bethe experienced this sense of wholeness at Los Alamos.

LOS ALAMOS

When one reads the reminiscences of those who had participated in the wartime project at Los Alamos, one has the sense that they look back upon this experience as if Los Alamos had been a utopia.[64] They had believed that they were in a frantic race to save the Western democracies. Their hope was that the atomic bomb would guarantee victory and secure a lasting peace. They knew they were involved in an enterprise which, if successful, would change the course of human affairs. In his presentation to the Gray Board, which reviewed his security clearance in 1954, Oppenheimer noted that Los Alamos

> was a remarkable community, inspired by a high sense of mission, of duty and of destiny, coherent, dedicated and remarkably selfless. There was plenty in the life of Los Alamos to cause irritation . . . but I have never known a group more understanding and more devoted to a common purpose, more willing to lay aside personal convenience and prestige, more understanding in the role that they were playing in their country's history. Time and time again we had in the technical work almost paralyzing crisis. Time and again the laboratory drew itself together and faced new problems and got on with the job. We worked by night and by day; and in the end the many jobs were done.[65]

In the Reith lectures that Oppenheimer delivered over the BBC in 1953, he recalled Los Alamos in more personal terms, in particular, his experiences "of the power and the comfort in even bleak undertakings, of common, concerted, co-operative life. Each one of us knows how much he has been transcended by the group of which he is a part; each one of us has felt the solace of other men's knowledge to stay his own ignorance, of other men's wisdom to stay his folly, of other men's courage to answer his doubts and weakness."[66]

To understand the character of the Los Alamos experience, we must remember that even though it operated under strict military and governmental surveillance, in many respects it was a democratic commune, a working democracy under trying, if not adverse conditions. This was true for the running of the "township," and it was true within the laboratory.[67] Los Alamos was unique in its enormous concentration of first-rate people who constantly gave proof of what could be accomplished by their working together on very circumscribed goals. It was an intense collaboration—in fact, a collaboration of unparalleled intensity. A cooperative task undertaken by outstanding people into which all threw themselves completely and singlemindedly, and to which all gave their ideas, experience and energy fully, freely, and selflessly. The intensity resulted in the total effort being much greater than the sum of its parts.

Though isolated—and perhaps because of its isolation—Los Alamos created this rare situation in the lives of individuals and communities in which they feel in touch with much more than themselves, in touch with almost everything. During the few years spent there, everyone felt whole. It was a merging of commitments, aspirations, inspirations, and talents that had been fragmented before. Not only did the participants feel whole as individuals—their moral, intellectual, and creative passions all being channeled into the task at hand—but an atmosphere of wholeness permeated the entire enterprise, transmuting it into a kind of magic and enshrining it in the minds of those who had been there. Oppenheimer, who was largely responsible for creating this sense of wholeness and maintaining it until the project was successfully completed, personified the integration of the multifacetedness of the enterprise: the theoretical and the experimental, the mundane and the idealistic, the individual, the community and the nation.[68]

When Oppenheimer left Los Alamos to return to Berkeley in the fall of 1945, the scientific staff presented him with a gift and read to him the following encomium:

> Let us . . . remember some of the things that are usually never mentioned. . . . He selected this place. Let us thank him for the fishing, hiking, skiing and for the New Mexico weather. He selected our collaborators. Let us thank him for the company we had, for the parties, and for the intellectual atmosphere which we will fully appreciate only when we again will be back at the universities, at faculty teas, and at student dances. He was our director. Let us thank him for the way he directed our work, for the many occasions where he was the eloquent spokesman for our thoughts. It was his acquaintance with every single little and big difficulty that helped us so much to overcome them. It was his spirit of scientific dignity that made us feel that we would be in the right place here. We drew much more satisfaction from our work than our consciences ought to have allowed us.[69]

Bethe was one of the speakers at the memorial for Oppenheimer that was held at the Institute for Advanced Study in the spring of 1967.

He opened his eulogy with the assertion that "J. Robert Oppenheimer did more than any other man to make American theoretical physics great." But it is of Los Alamos that Bethe talked most movingly:

> Los Alamos might have succeeded without him, but certainly only with much greater strain, less enthusiasm, and less speed. As it was, it was an unforgettable experience for all the members of the laboratory. There were other wartime laboratories of high achievement. . . . But I never observed in any one of these other groups quite the spirit of belonging together, quite the urge to reminisce about the days of the laboratory, quite the feeling that this was really the great time of their lives.
>
> That this was true of Los Alamos was mainly due to Oppenheimer. He was a leader. It was clear to all of us, whenever he spoke, that he knew everything that was important to know about the technical problems of the laboratory, and he somehow had it well organized in his head. But he was not domineering, he never dictated what should be done. He brought out the best in all of us, like a good host with his guests. And clearly because he did his job very well, in a manner all could see, we all strove to do our job as best we could.[70]
>
> One of the factors contributing to the success of the laboratory was its democratic organization. The governing board, where questions of general and technical laboratory policy were discussed, consisted of division leaders (about eight of them). The coordinating council included all the group leaders, about 50 in number, and kept all of them informed on the most important technical progress and problems of the various groups in the laboratory. All scientists having a B.A. degree were admitted to the colloquium in which specialized talks about laboratory problems were given. Each of these three assemblies met once a week. In this manner everybody in the laboratory felt a part of the whole and felt that he should contribute to the success of the program. Very often a problem discussed in one of these meetings would intrigue a scientist in a completely different branch of the laboratory, and he would come up with unexpected solutions.
>
> This free interchange of information was entirely contrary to the organization of the Manhattan District as a whole. . . . Oppenheimer had to fight hard for free discussion among all qualified members of the laboratory. But the free flow of information and discussion, together with Oppenheimer's personality, kept morale at its highest throughout the war. Los Alamos has been an example for big accelerator laboratories ever since,

and although they are concerned with very different scientific problems, Brookhaven and CERN and many other places have gained much of their spirit from wartime Los Alamos.

And one of these "other places" was the Newman Laboratory at Cornell.

It should be emphasized that the sense of wholeness and unity that permeated Los Alamos was the result of the intense collaborative effort on an engineering project *whose aim was sanctioned by all those working on it*. It was a project driven by political and ideological commitments, cemented by social organization, that was headed by a charismatic leader who galvanized the energy and devotion of the participants. At Oppenheimer's trial in 1954, Bethe stressed that "the success of Los Alamos rested largely on teamwork and the leadership of its director. . . . I have never met anyone who performed the functions [of director] as brilliantly as Oppenheimer."[71]

One further aspect of the Los Alamos enterprise merits comment. The seeming glory of the achievement was so great that no one took anything away from anyone else. Each participant—with the exception of Oppenheimer—got equal credit. Additionally, Oppenheimer's directorship of Los Alamos and the crucial contributions made by the theoretical division headed by Bethe earned theorists new respect and authority.

As it had been for Oppenheimer, Los Alamos was a turning point in Bethe's life. His stay there was a period of enormous productivity. The war had brought out his impressive talents in applied physics, and these were consolidated there. At Los Alamos he successfully directed the activities of the largest numbers of theorists ever assembled—and among these were the very best of them including Peierls, Serber, Weisskopf, and Feynman. It was there that he first encountered Feynman, and it was there that their deeply felt friendship developed. At Los Alamos, Bethe partook in an enormously stimulating social life: on most weekends there would be get-togethers with friends and hikes to the nearby mountains. Los Alamos also brought him closer to his wife, with whom he had shared the nature of the activities in which Los Alamos was engaged. Both of Bethe's children's were born at Los Alamos.

BETHE AND OPPENHEIMER: THEIR ENTANGLEMENT

Although they had met previously at some of the Washington, D.C., meetings of the American Physical Society, the first substantial interaction between Bethe and Oppenheimer occurred in the summer of

1940 when Bethe had accepted an invitation to teach at the summer school at Stanford University. On the way to California he stopped in Seattle to present a paper on cosmic rays at the meeting of the American Physical Society that was being held there from June 18 to 21. The afternoon session on June 20 was devoted to an informal symposium on theoretical physics. The opening talk was by Oppenheimer, who spoke on "The Present Crisis in the Quantum Theory of Fields"; Bethe, G. M. Volkoff, and Hartland Snyder were the other participants on the panel. That same evening, both Bethe and Oppenheimer were invited to a small evening party of about ten people at the house of Edwin Uehling, one of Oppenheimer's former postdoctoral students. All present were deeply depressed by the turn of events in the war—France had just fallen, England was in imminent danger, and the Hitler-Stalin pact was still intact. Everyone present was terribly anxious about the future. Bethe recalls Oppenheimer addressing the group and stating in very simple words: "This is a time when the whole of western civilization is at stake. France, one of the great exponents of western civilization, has fallen, and we must see to it that Britain and the United States don't fall as well. We have to defend western values against the Nazis. And because of the Molotov–von Ribbentrop pact we can have no truck with the Communists."[72] To Bethe, the little speech "showed such a deep understanding of the situation. And at the same time he had this wonderful facility with words. It was most impressive." This was the first time that Bethe had an intense, intimate discussion with Oppenheimer.[73] Bethe believes that it may have been "the first occasion in which Oppenheimer talked about political matters, not from the standpoint of the left, but from the standpoint of the West."[74] After June 1940, Oppenheimer became particularly intent on using his science in his stand against Nazi Germany and completely repudiated his leftist proclivities—sentiments that had been very strong before. His earlier political views can be gauged from a letter that Felix Bloch had written to Rabi in the fall of 1938: "Oppje[75] is fine and sends you his greetings; honestly, I don't think you wore him out but at least he does not praise Russia too loudly any more which is already some progress."[76]

The Stalinist purges of 1936–38, the Moscow trials, and the report of the Commission of Inquiry that Dewey had headed to evaluate the charges made against Trotsky had made a deep impression on many intellectuals who had been sympathetic with the Soviet Union.[77] Dewey had visited the Soviet Union in 1928, and at the time he was optimistic about the prospects of freedom there. But he became bitterly disillusioned by what he had learned chairing the Commission of Inquiry. He had gone to Mexico to meet with Trotsky and thereafter commented: "The great lesson for all American radicals and for all sympathizers with the USSR is that they must go back and consider the whole question of the means of bringing about social changes and truly democratic methods of approach to social progress."[78]

Rabi had always been wary of communism and of the Bolshevik experiment in Russia. He agreed with Dewey, his colleague at Columbia, who had asserted: "The dictatorship of the proletariat has led and, I am convinced always must lead to a dictatorship over the proletariat and over the party. I see no reason to believe that something similar would not happen in every country in which an attempt is made to establish a Communist government."[79]

Rabi and Oppenheimer had become friends in the late 1920s when both were in Germany. The friendship grew during the 1930s and they came to deeply respect one another. Rabi admired Oppenheimer's off-scale abilities but was not intimidated by him; Rabi was also aware of the frailty of Oppenheimer's character. Oppenheimer appreciated Rabi's political acumen, his shrewdness and savvy, and, above all, his forthrightness.

As Bloch's letter indicates, Rabi's political views were having an impact on Oppenheimer. In the fall of 1936, Oppenheimer had become deeply involved in "left wing" causes, sensitized by his relationship with Jean Tatlock, the daughter of a professor of literature at Berkeley, and by his friendship with his brother, Frank. As he recalled before the Gray Board in 1954:

> I should not give the impression that it was wholly because of Jean Tatlock that I made left-wing friends, or felt sympathy for causes which hitherto would have seemed so remote from me, like the Loyalist cause in Spain, and the migratory workers. . . . I liked the new sense of companionship, and at the time felt that I was coming to be part of the life of my time and country.[80]

The fall of France was a watershed. When the opportunity presented itself to work on the uranium project, Oppenheimer eagerly accepted. He worked on Lawrence's project that had as its goal the electromagnetic separation of U^{235} from U^{238} in naturally occurring uranium ores.

Early in 1942 Oppenheimer was assigned the responsibility for the investigation of fast neutron fission and for the design of an atomic bomb. The concept of a fission bomb and its feasibility had first been analyzed by Otto Frisch and Rudolf Peierls and their associates in England.[81] It became the focus of a thorough study by theorists working with Oppenheimer in Berkeley during the summer of 1942. Bethe and Teller were asked by Oppenheimer to join the workshop which was concerned with the theoretical design of an atomic bomb and the estimation of its efficiency.[82] Thus they were all together at this Berkeley conference to make recommendations regarding the feasibility of an A-bomb. It became clear early on that a U^{235} fission bomb was indeed achievable, and a good part of the summer became devoted to an inquiry whether deuterium could be made to "burn" when triggered by a fission bomb. It was at this workshop that "the Super" got hatched. Teller later recalled:

> It is hard to describe the intensity and the fascination of the discussion that followed [Teller's question of whether deuterium could be exploded.] . . . The experience proved perhaps even more challenging than the previous discussion [at the 1938 Washington conference] about the interior of the sun. Here we were not bound by the known conditions in a given star but were free within considerable limits to choose our own conditions. We were embarking on astrophysical engineering.[83]

Perhaps one of the characteristic differences between Bethe and Teller is that Bethe never became an "astrophysical engineer." The results obtained by this study group bolstered the conclusion that Peierls and Frisch had obtained. It refined their estimate of the amount of U^{235} necessary and gave a better estimate of the efficiency of such a weapon. In the spring of 1943, the Los Alamos Laboratory was established to develop and assemble an atomic bomb.

Bethe noted that

> there was a tremendous change in Oppenheimer from 1940 to 1942, and especially in 1943. In 1940 he was confused, he mumbled, he certainly wouldn't have given anybody any orders. He was taken very much with esoteric problems. He had done very good scientific work, . . . but he was attracted by problems beyond the capacity of anybody to solve, including his. And, so, when I listened to his seminar in 1940, I was not impressed, because those people talked a lot in very learned ways about things none of them understood. In 1942 the new personality had gelled. He was much more decisive. He ran this group on the theory of nuclear methods very firmly. [During the summer of 1942] he ran our group of consultants well. It didn't need much running, but it was clear that he was in command. But the real development was between then and '43. Then he really came into his own, and he obviously had always wanted to accomplish something definite, something outstanding. And Berkeley and Caltech had not given him that opportunity.

Oppenheimer chose Bethe to head its theoretical physics division,[84] the position that Teller had aspired to. This was probably the trigger for the subsequent strained relations between Teller and Oppenheimer at Los Alamos. Bethe thus became a member of the Governing Board, which was made up of division leaders and key administrative and liaison personnel, and of the Coordinating Committee, made up of all the group leaders; both bodies advised Oppenheimer on overall policy matters. Bethe strongly endorsed the recommendation to have a weekly colloquium open to all scientifically qualified personnel at which presentations of the work of the different divisions would be presented. The colloquium became the vehicle by which staff members were informed of the general developments at the laboratory and played an important role not only in keeping morale and momentum up, but as a forum where ideas and expertise were exchanged. It became the most important demonstration and exhibition of the solidarity of the enterprise. The colloquium helps explain the moral response of the Los Alamos physicists to the bombing of Hiroshima and Nagasaki. The colloquium had given everyone the conviction that they were an integral part of the project. Everyone in attendance had been privy to all aspects of the project and had full knowledge of the goals of the enterprise. Full knowledge in turn engenders responsibility.

Although at the professional level the interaction between Bethe and Oppenheimer was very close, the Bethes and the Oppenheimers did not see that much of one another at the social level, partly because Rose Bethe was wary of Kitty Oppenheimer.

After the war, Bethe and Oppenheimer were in frequent telephone contact and saw each other at American Physical Society meetings in New York and Washington. In the spring of 1946, at Bethe's invitation, Oppenheimer delivered the Messenger lectures at Cornell University,[85] and Oppenheimer reciprocated by inviting Bethe to be the Hitchcock professor at UC–Berkeley that winter.

When Oppenheimer resigned from Berkeley in the spring of 1947 to accept the directorship of the Institute for Advanced Study, Raymond Birge, the chair of Berkeley's physics department, declared that "the loss of Professor Oppenheimer must be considered the greatest blow ever suffered by the department, and I cannot help but consider that my own failure to persuade him to remain here is the greatest failure of my life. The whole situation is a direct result of the war, and of the conditions brought about by the war." The Berkeley physics department had in fact never lost a distinguished member of its staff since the founding of the university "except by death or retirement."[86]

Oppenheimer's resignation had not come suddenly. He had pondered the matter for several months and had conferred at length with Birge. The issue of who could replace him should he leave was undoubtedly discussed. Upon Oppenheimer's resignation the position was offered to Bethe,[87] but Bethe did not accept.

Bethe and Oppenheimer both participated in the influential series of workshops on theoretical physics held at Shelter Island, Pocono, and Oldstone in 1947, 1948, and 1949, respectively. These conferences resulted in the formulation of renormalization theory and the two-meson hypothesis.[88] Duncan McInnes, a physical chemist at the Rockefeller Institute, and K. K. Darrow, the secretary of the American Physical Society, had organized the Shelter Island Conference and Darrow chaired it. McInnes's impression of the gathering gives a revealing account of Bethe and Oppenheimer after the war. McInnes recorded in his diary that "it was immediately evident that Oppenheimer was the moving spirit of the affair." Oppenheimer was the dominant personality and "in absolute charge." When, at Darrow's insistence, the evening sessions were terminated at 9:00 P.M., the appointed hour, Oppenheimer adjourned the normal sessions to informal meetings that lasted late into the night. Darrow, in his diary, gave the following sketch of Oppenheimer:

> As the conference went on the ascendency of Oppenheimer became more evident—the analysis (often caustic) of nearly every argument, that magnificent English never marred by hesitation or groping for words (I never heard "catharsis" used in a discourse on [physics], or the clever word "mesoniferous" which is probably O's invention), the dry humor, the perpetually-recurring comment that one idea or another (incl. some of his own) was certainly wrong, and the respect with which he was heard. Next most impressive was Bethe, who on two or three occasions bore out his reputation for hard & thorough work, as in analyzing data on cosmic rays variously obtained. (An amusing interchange in which X—I've forgotten who, Teller I think— had put a math'l argument on the board; Y said "is there not a logarithm?"; X replied "when I do it decently the logarithm will be there"; Bethe said "When it is done *really* decently, there is no logarithm" [laughter!].)[89]

Bethe and Oppenheimer's professional interactions also involved national security matters, and more specifically nuclear weaponry. After the detonation of the first Russian atomic bomb, Joe 1, on August 29, 1949, and its detection a few days later by the radioactivity in air samples collected by the air force monitoring flights,[90] the issue of whether to meet this threat by developing a hydrogen bomb became the focus of intense discussions. Edward Teller was a strong proponent of developing such a weapon. As we shall see in chapter 5, Bethe and Oppenheimer took a different view of the matter. Bethe and Teller visited Princeton to discuss with Oppenheimer, who was chairing the General Advisory Committee (GAC) of the AEC, the feasibility of a "Super."[91] Shortly thereafter, Bethe informed Teller that he would not work on such a project—and subsequently Teller accused Oppenheimer of influencing Bethe in arriving at his decision. It was actually discussions with Victor Weisskopf, George Placzek, and with Rose, his wife, that led Bethe to take the stand he did. The recommendation that Oppenheimer and the GAC made to the AEC—against a crash program to develop a fusion bomb—was one of the factors that led to the revocation of his clearance in 1953. Bethe was one of Oppenheimer's staunchest supporters at his trial in 1954.

The continued esteem that Oppenheimer had for Bethe is manifested by the fact that the Institute for Advanced Study offered Bethe a professorship in 1953, which he declined. And indicative of the bond that had formed between these two men over the years is the handwritten note to Oppenheimer that Bethe appended to the form letter he had sent to the several hundred people who had come to Ithaca or sent him letters on the occasion of the celebration of his sixtieth birthday in October 1966:

> Thanks for your especially warm telegram. It was very good to see you 2 weeks ago. Your words express, better than I can, what I feel for you—admiration, affection, enduring gratitude and friendship.
>
> As ever,
> Hans

Oppenheimer at that time was terminally ill—and was unable to attend. He had sent Bethe the following telegram: "As you know my thoughts will be with you especially on your day today. With admiration, affection and enduring gratitude and friendship."[92] Oppenheimer had been valiantly battling throat cancer for over a year, and the cancer had metastasized. He could talk about it, and his eventual death, "as lucidly as he could discuss a conclusion in physics."[93] He died on February 18, 1967. Shortly thereafter, Bethe wrote a lengthy "Biographical Memoir" of Oppenheimer for the Royal Society of London, of which both were Fellows.[94]

I next turn to two incidents during the early Cold War to highlight the way Oppenheimer and Bethe responded in the defense of colleagues and of civil liberties at the time when the foundations of the national security state were being laid in the United States. Oppenheimer's actions in the Bernard Peters case represent the nadir in his political activities during the McCarthy era and reveal flaws in his character. His actions in that case help explain why for all their admiration and respect for the man, the American physics community could not subsequently cast him as the persona embodying its commitments and values. Bethe acted commendably in the Morrison case. Nonetheless, he too gave evidence of the enormous strains that McCarthyism generated as a result of McCarthy's misdeeds. It was Robert Wilson and Dale Corson, Bethe's colleagues in the Newman Laboratory at Cornell, who were the exemplary figures as guardians of democracy and academic freedom. But what emerges from the account is Bethe's deep sense of integrity, a quality nurtured and strengthened by the community he had helped form and of which he was an integral part.

4.

THE CHALLENGE OF McCARTHYISM

> In these times, when our American way of life is so much in the balance, often suffering as much from overzealous defenders as from cruel attack from without, the scientist can hardly remain silent.
>
> —I. I. Rabi, at opening exercises, Columbia University, September 26, 1950

THE BERNARD PETERS CASE

By 1949 the Cold War had become frigid and the national mood in the United States was one of "cold fear." West Berlin was being blockaded and most Americans came to believe that the goal of the Soviet Union was to impose its totalitarian system on the entire world, as it had on Eastern Europe. An anti-Communist hysteria and a concomitant obsession with atomic secrets were gripping the country. The rampant fear of Communism that had started with the unraveling of the wartime alliance with the Soviet Union had led to a policy of containment that had been drafted by George F. Kennan.[1] The fears engendered by the perceived Soviet threat unleashed a wave of political repression that lasted through the 1950s. The blacklisting of Hollywood actors, directors, and writers, the Hiss and the Rosenberg cases, J. Edgar Hoover's anti-Communist obsession, Senator Joseph McCarthy's career, were all facets of this anti-Communist campaign. This crusade to purge the presumed insidious effects of the spread of Communist ideology has sometimes been characterized as "paranoid," but the fact is that a real threat existed. With the opening of Soviet-era state archives, explicit proof of espionage by the USSR became available: the code names and salary receipts of the agents and copies of the sensitive materials on the Manhattan Project and on American diplomatic strategy that were given to the Soviets can be found there.[2] Julius Rosenberg and Alger Hiss were evidently guilty of the charges brought against them. There were, in fact, indications of subversion by Communist party members at the time. Thus, in 1946 the Canadian government announced the arrest of twenty-two persons charged with illegally passing information to Soviet officials. It also declared that it was very likely that the NKVD, the Soviet intelligence agency, had established espionage rings in the United States and

Great Britain. However, the danger was exaggerated and the responses to it—in particular, the loyalty oaths that were introduced to
ferret out Communist influence in the federal government, in state
governments, and in school systems—were out of proportion to the
reality of the situation. The harm that was inflicted on the United
States from Soviet-sponsored spies did not justify McCarthy's witch
hunts and wave of terror. Not only did he wreck individual lives, but
he silenced alternative political discourse, thus undermining the
foundations of American democracy.

To understand the McCarthy era, we must recognize that it constituted a continuation of the struggle that had been waged between
conservatives and liberals during the 1930s.[3] The election of Franklin
D. Roosevelt had been an important victory for the liberal cause. The
conservative forces capitalized on the Soviet threat as a means to regain control in the political arena. An alliance developed among conservative members of Congress, conservative elements within the executive branch, and the FBI, and it would not be an exaggeration
to characterize the activities of the American government that were
directed against scientists as those of a police state. Scientists, and
physicists in particular, were vulnerable because they were seen as
the possessors and generators of valuable state secrets who also pro-
fessed loyalty to an international community of fellow investigators.
Oppenheimer came to be seen as the paradigm of the scientist. But
he was a scientist whose past political activities raised serious doubts
in the minds of the most extreme elements in the conservative alliance as to his trustworthiness.

Until Senator Joseph McCarthy took center stage in the early 1950s,
the "Red Scare" was abetted by the hearings and attendant leaks of
the House of Representative's Committee on Un-American Activities
(HUAC).[4] Most of the representatives on HUAC were rabidly anti–New
Deal, committed to discredit the whole program of liberalism. During
the Eighty-first Congress, HUAC was chaired by John S. Wood, a
southern Democrat from Georgia. Sitting on the committee was a
young Republican from California, Richard M. Nixon, who had won
his seat in Congress in 1946 with a "dirty" campaign against the in-
cumbent Democrat, Jerry Voorhis. The most vociferous and strident
member of HUAC was C. Parnell Thomas, a conservative Republican
from New Jersey. He had been chairman of HUAC during the Eightieth
Congress and had catapulted the Committee to national prominence
by initiating a program "to expose and ferret out the Communists and
Communist sympathizers in the Federal government, . . . in some of
the most vital unions of American labor, . . . in Hollywood, . . . in education, and to investigate those groups and movements which are

trying to dissipate [*sic!*] our atomic bomb knowledge for the benefit of a foreign power."[5]

The headlines on the front page articles of the *New York Times* for June 9, 1949, give a good indication of the temper of the times. They reported as follows:

ACHESON PROPOSES BIG 4 SET DEADLINES ON BLOCKADE TALK
ISOTOPE SHIPMENT TO NORSKE STIRS ROW
HISS TO TAKE STAND; CHAMBERS SAYS HE HAD 5 "SOURCES"
GATES ADMITS REDS TAUGHT REVOLUTION
FILM "COMMUNISTS" LISTED IN FBI FILES IN COPLON SPY CASE
EISENHOWER AND CONANT IN GROUP BARRING COMMUNISTS AS TEACHERS

On page 4, a full-column article told that "Hundred Named as Red Appeasers. California's Tenney Committee Lists Actors, Musicians and Others as 'Line' Followers." Among those cited besides actor Charlie Chaplin, playwright Lillian Hellman, and former vice-president Henry Wallace were Nobel laureate Thomas Mann and Haakon Chevalier, a professor of French literature at Berkeley and a friend of J. Robert Oppenheimer.

The headline of the major article on page 2 read: "3 FACING CONTEMPT; SILENT ON IDEOLOGY." The article reported that HUAC had ordered its counsel to draw up a contempt citation against Steve Nelson for refusing to answer any questions, including those of party affiliations. Nelson, the head of the Communist party in Alameda County, had fought in the Spanish Civil War alongside Joe Dallet, Kitty Oppenheimer's first husband.[6] In the late 1930s, he had been a regular visitor to the Berkeley campus of the University of California, promoting causes with which many of the faculty, including Oppenheimer, and their students had sympathized.[7] HUAC had accused Nelson of organizing a small cell of Communist party members at the Radiation Laboratory during World War II, of receiving from one member of this cell, scientist X, highly secret data concerning the atomic bomb project, and of having transmitted to the Soviet vice consul in San Francisco the data that scientist X had given him.[8]

Since April 22, 1949, HUAC had been holding hearings regarding "Communists' infiltration of Radiation Laboratory and atomic bomb project at the University of California, Berkeley" in an attempt to "obtain further enlightenment concerning the scientist X case." These hearings were the continuation of an unproductive HUAC investigation of atomic espionage of a year earlier. The Committee had claimed to have discovered—in closed executive sessions[9]—security violations at the Radiation Laboratory involving Frank Oppenheimer,

Robert's brother, and four former students of Robert Oppenheimer: David Bohm, Irving David Fox, Rossi Lomanitz, and Joseph Weinberg.[10] One of these was presumably scientist X.[11] In point of fact, all the HUAC hearings had been aimed at Oppenheimer, who by 1949 had enormous influence in government circles but held views that many on Capitol Hill and in the air force thought to be too pacific.[12]

The questioning of Nelson by HUAC was described by the *New York Times* as having been an "arm-waving, gavel-pounding session in which members and witnesses shouted at each other." The article went on to report as follows:

> After Mr. Nelson stepped down, the committee heard Dr. Bernard Peters, physicist at the University of Rochester, in a closed session. Dr. Peters worked at the radiation laboratory at Berkeley during the war.
>
> Representative Wood said that Dr. Peters answered all questions "fully and freely." He said he would not be recalled. Dr. Peters said he replied "no" when asked if he was a Communist. He told the Committee he did not know Mr. Nelson, but did know Joseph Weinberg, who also worked at Berkeley and had been questioned secretly by the committee.[13]

Actually, two days earlier, on the morning of Tuesday, June 7, Oppenheimer had testified before HUAC in a closed executive session about his former students—Weinberg, Lomanitz, and Peters. He had been asked by the committee "for his assistance" in their probe of the alleged Communist cell in the Radiation Laboratory.[14] At the start of the hearing, HUAC counsel Frank S. Travenner assured Oppenheimer that the committee was "not seeking to embarrass" him since his "record of loyalty had been vouched for by General Groves."[15] Indeed, Oppenheimer was seen as the exemplary scientist-citizen. His image in the national limelight can be gauged by *Time* magazine's coverage of his testimony before the Joint Atomic Energy Committee on June 13, 1949.[16] The Joint Committee, at the time, was investigating the administration of the operations of the Atomic Energy Commission at the request of its chairman, David Lilienthal, after the charge of "incredible mismanagement" had been leveled against him by Senator Bourke Hickenlooper, who was trying to get Lilienthal to resign as chairman of the Commission. *Time* reported that

> boyish-looking Robert Oppenheimer stole the show. . . . For 2 hours the cropped-haired scientist set forth the intricacies of atomic science, gave sure, rapid-fire answers to polite questions and punched gaping holes in Iowa Senator Bourke Hickenloop-

er's foundering campaign. . . . Like wide-eyed students en-thralled by their favorite professor, committee members thanked their witness and apologized for keeping him so long.[17]

On June 15, 1949—presumably on the basis of information that C. Parnell Thomas had divulged to its Washington bureau about the closed-door testimony of Oppenheimer before HUAC on June 7—the *Rochester Times-Union* carried the headline: "DR. OPPENHEIMER ONCE TERMED PETERS 'QUITE RED'." The lead story stated that

Dr. J. Robert Oppenheimer, wartime director of the atom Labo-ratory at Los Alamos, N.M. recently testified that he once termed Dr. Bernard Peters of the University of Rochester "a dan-gerous man and quite Red." . . . [H]e had known the scientist as a graduate student in the physics department in the late 1930's.

Said Dr. Oppenheimer:

"Dr. Peters was, I think, a German National. He was a mem-ber of the German National Communist Party. He was impris-oned by the Nazis and escaped by a miracle. He came to this country. I know nothing of his early period in this country. He arrived in California, and violently denounced the Communist Party as being a 'do-nothing party.'"

Dr. Oppenheimer said he told Major DeSylva [*sic*] he believed Dr. Peters' background was filled with incidents that would point toward "direct action."

Asked to explain this point, Oppenheimer explained:

"Incidents in Germany where he fought street battles against the National Socialists on account of Communists; being placed in a concentration camp; escaping by guile. It seemed to me those were past incidents not pointing to temperance."

Questioned specifically on his reference to "direct action," Dr. Oppenheimer said of Dr. Peters:

"I think I suggested his attack on the Communist Party as being too constitutional and conciliatory an organization, not sufficiently dedicated to the overthrow of the Government by force and violence."

Asked the source of his information that Dr. Peters had been a member of the Communist Party in Germany, Dr. Oppenhei-mer replied:

"It was well known. Among other things, he told me."

Dr. Oppenheimer said he could "affirm that there is no con-nection between his (Peters') work and any application of atomic energy that falls in the jurisdiction of the (Atomic En-ergy) Commission. . . . I would believe that if Dr. Peters could teach what he knows to a young man capable of learning it, the country would be better off, because if Dr. Peters cannot be employed by the War Department, at least, the young man could be employed by the War Department."

Bernard Peters had been a student of Oppenheimer's. He was born in the German city of Posen (now Poznan, Poland) in 1910 and moved to Freiburg in the Rhine Valley with his family two years later.[18] In 1932 he went to the Technical University in Munich to study electrical engineering. After Hitler came to power, Peters became an active opponent of the Nazi regime, was arrested and sent to Dachau, then to a prison camp. He escaped the camp after three months and traveled by night on a bicycle through the Alps to Italy. There he joined his girlfriend, Hannah Lilien, who had gone to Padua to study medicine. From Italy, Peters went to England, where he stayed until 1934 when he obtained a visa to the United States.[19] Hannah soon joined him and they were married in New York. To support them and to allow Hannah to finish her medical studies, Peters worked in an import firm. In 1938, after Hannah received her medical degree, they moved to the San Francisco Bay Area. She obtained a research position at the Stanford Medical School, and he worked as a longshoreman in San Francisco. He met Oppenheimer at a social gathering and Oppenheimer encouraged him to come to Berkeley as a graduate student in physics.[20] He took a course with Oppenheimer who, taken with Peters's talents in theoretical physics, got him admitted to the graduate program in physics even though Peters did not have the requisite undergraduate accreditation. Peters became the note taker for Oppenheimer's quantum mechanics course and in this capacity got to see Oppenheimer regularly and frequently.[21] Moreover, since Peters chose to do a dissertation with Oppenheimer, he became a member of Oppenheimer's circle of prize graduate students and postdocs. Philip Morrison, who was then a graduate student in the physics department at Berkeley doing a dissertation with Oppenheimer, got to know Peters well. He recalled that Peters was more realistic about the state of the world than the other students.[22] He

> seemed a little different from most of us [the dozen young graduate students around Robert Oppenheimer], more mature, marked with a special seriousness and intensity, . . . his experience went far beyond ours. . . . He had seen and felt the barbarous darkness that mantled Nazi Germany, [and] had worked among longshoremen in San Francisco Bay. In contrast, we were . . . largely innocent save by hearsay and our eager reading of most of life beyond the campus and overseas.[23]

Also, because Peters was somewhat older and more cosmopolitan than the other students, a personal friendship developed between him and Oppenheimer.[24] After Oppenheimer married Kitty, the Peterses and the Oppenheimers saw "quite a lot" of one another socially. In late 1942, after he had been named director of Los Alamos, Oppenheimer raised with Peters "the question of whether he would come [to Los Alamos]. The fact that he was the right kind of physicist and that she was a doctor and we were short of doctors made this an attractive deal." But the Peterses decided not to go,[25] and Bernard Peters stayed at the Radiation Laboratory in Berkeley.[26]

A few months later, during the questioning to obtain his security clearance, Oppenheimer evidently averred that his former students David Bohm, Joseph Weinberg, and Bernard Peters had been Communists. He stated that, in 1933, as a student in Munich, Peters had taken part in the anti-Nazi struggle "as a Communist."[27]

In early 1943, when Oppenheimer was interrogated by Peer de Silva, the officer in charge of security matters at Los Alamos, Bernard Peters's name came up once again. De Silva had asked: "Here are four names, Bohm, Weinberg, . . . and Peters;[28] which of these would you regard as the most likely to be dangerous?" and Oppenheimer had answered "Peters." A possible, charitable interpretation of Oppenheimer's reply is that, given Peters's brilliance, his past experiences, his determination, stamina, and tenacity—as evidenced by what he accomplished in physics in his few years at Berkeley—Peters was *potentially* more dangerous than the others whom de Silva had named. But Oppenheimer must thereafter have realized that his answer raised questions about his judgment in having invited Peters to join the project a few months earlier. In January 1944, on the train that was taking them both back to Los Alamos from a trip to Berkeley, Oppenheimer had another conversation with de Silva. According to de Silva, during the interchange Oppenheimer brought up the "situation in Berkeley . . . and touched on the subject of what persons he thought dangerous were in his opinion truly dangerous." De Silva, in a memorandum dated January 6, 1944, penned upon his arrival in Los Alamos, recorded that

[Oppenheimer] named David Joseph Bohm and Bernard Peters as being [truly dangerous]. Oppenheimer stated, however, that somehow he did not believe that Bohm's temperament and personality were those of a dangerous person and implied that his dangerousness lay in the possibility of being influenced by others. Peters, on the other hand, he described as a 'crazy person' one of whose actions would be unpredictable. He described Peters as being "quite a red" and stated that his background was filled with incidents which indicated his tendency toward direct action.[29]

Evidently, some member of HUAC or its staff had leaked to the *Rochester Times-Union's* Washington bureau not only a summary of Oppenheimer's testimony at the closed session of HUAC, but also some of the information in his FBI file, in particular, the de Silva memorandum.

In the column next to the article that described Oppenheimer's testimony before HUAC, the *Times-Union* printed an interview with Peters. In it, Peters denied ever having been in the Communist party or ever having attended Communist meetings except in Germany, and then not as a party member. He never fought street battles against the Nazis, "but wished he had." As a student in Germany, he was in one anti-Nazi demonstration where two people got hurt. He never told Dr. Oppenheimer that he had been a member of the Communist party "because he never was"—but he did tell him that "he greatly admired the spirited fight they put up against the Nazis, especially in Europe after the Nazi occupation." Nor had he ever advocated the overthrow of any democratically elected government by force, though he would advocate the overthrow of a Fascist regime like Franco's. He emphasized that he never made a secret of his political beliefs, which he characterized as "not orthodox," as he was in favor of socialism as an economic system. And to undermine the credibility of Oppenheimer's testimony, Peters asked: "If the intelligence department of the Manhattan project did believe Oppenheimer's information, why was I kept on the project till 1946?"

The article in the *Times-Union* jeopardized Peters's position at Rochester, even though he was considered an unusually able cosmic ray physicist and had earned the respect and friendship of his colleagues. At the time, he was actively engaged in analyzing nuclear reactions induced by cosmic rays in photographic emulsions sent to the stratosphere in high altitude "Skyhook" balloons. In 1947, in a collaboration with a team of cosmic ray physicists at the University of Minnesota that included Frank Oppenheimer, the particle tracks in Helmut Bradt and Peters's emulsions gave evidence for heavy nuclei as a component of the primary cosmic radiation. Furthermore, Bradt and Peters found that the chemical composition of primary cosmic rays is very similar to the "universal" chemical composition of stars and galaxies. Their presence, in addition to protons and alpha-particles, the most abundant component, thus pointed to the galactic origin of cosmic rays and made cosmic ray physics part of astrophysics rather than cosmology.[30] Later in 1949, Peters would report some startling new findings, namely the discovery of the neutral pion and of new "heavy mesons."[31]

After hearing of Oppenheimer's testimony Peters wrote him an angry letter to find out whether he indeed had made the statements attributed to him, and if so on what basis. Understandably, Peters was appalled and couldn't understand how Oppenheimer could have done him such harm.[32] Moreover, Peters had stopped in Princeton on June 9 on his way home from Washington to talk to Oppenheimer. During their conversation, Oppenheimer had told Peters that he had testified before HUAC two days earlier, but he did not reveal that Peters had been discussed extensively. According to a letter written by Condon to his wife on June 23, 1949, which recounted what Peters had reported to him about his visit in Princeton, Oppenheimer had declared: "God guided their questions so that I did not say anything derogatory." Furthermore, "He was extremely cordial to Peters and insisted that Peters look him up in Berkeley when both would be out there several weeks hence. All was sweetness and love."[33]

The news of Oppenheimer's HUAC testimony spread rapidly within the physics community. The reaction was generally one of great disillusionment—for Oppenheimer had a special standing among physicists both as an outstanding scientist and an admirable citizen. The community had expected greater integrity from a teacher in his dealings regarding his students—especially when the issues involved were political and concerned the academic freedom of a university professor.[34] The letter written by Victor Weisskopf to Oppenheimer captures the dismay of the community:[35]

Dear Robert,
    I was very unhappy when I read the newspaper report on your hearing about Bernard Peters a few days ago. Let me say right away that I do not believe for a moment that you really have said the things which are ascribed to you in this report. They must have misunderstood completely your statement and your aims.
    However, this report is the only known record of this hearing until you have made some statement to the contrary.
    I know Peters and I know his opinions and his bias. Let me add that I don't like him very much because of his intransigence and his lack of humor and human understanding.
    He is now in great danger of losing his job because of this report. His case is far more important than seen from the personal angle.   If Peters loses his job because of the statement about his

political leanings made by *you* (whether you really made these statements or whether they were only ascribed to you is irrelevant) we are all losing something that is irreparable. Namely confidence in *you.* I beg you therefore, Robert, *set this record straight* and do what is in your power to prevent Peters' dismissal, even if you have to pay for it by losing reputation somewhere else.

Whatever Peters' political ideas are, his dismissal is unjustified and would be an exact parallel to what is going on in Russia.

I know he is not the first scientist who has been fired because of political leanings. He would be the first one, however, that is fired on the basis of information from another scientist, and what is more, from you[36] whom so many regard as our representative in the best sense of the word.

I am sure you can justify every word you said to the committee. But it is the published record and its effect that counts, and nobody can read this unhappy record without being thoroughly shocked.

Please do whatever you can to repair this! There is so much at stake.[37]

Ps. Please excuse my emotions. After reading this letter I first decided to rewrite it. But on second thought, you may just as well know how I felt.[38]

One aspect of Weisskopf's poignant letter should be noted. Weisskopf was well aware that scientists in the past had encountered difficulties because of their ideological beliefs and their metaphysical commitments. But in those cases—and Galileo springs to mind—it was not another scientist who had pointed the finger at them.[39] In Weisskopf's eyes, what Oppenheimer was guilty of was transgressing the ethos of the scientific community—a community held together by mutual trust—and Oppenheimer had violated that trust. In fact, he had shattered more than that, since in Weisskopf's eyes Oppenheimer was the *charismatic* leader of that community. Hence his lament: "We are all losing something that is irreparable. Namely confidence in *you.*"

Weisskopf had sent his letter to Oppenheimer from Boulder, Colorado, where he was visiting David Hawkins, a friend from Los Alamos, who would later have serious difficulties with Jenner's Subcommittee on Internal Security. Weisskopf had intended to go from Boulder to Idaho Springs to attend a cosmic ray conference being held there in late June 1949; but he found the Hawkinses and the Rockies so pleasant that he stayed in Boulder. Both Peters and Frank Oppenheimer participated in the Idaho Springs meeting.[40] Their plight became the subject of most of the discussions outside the lecture room, and their predicament weighed heavily on the minds of the one hundred or so physicists in attendance. Bethe, Condon, Millikan, and Teller were at the conference, and they had been deeply disturbed by what they had read of Oppenheimer's testimony in the Rochester newspaper. Like Weisskopf, Bethe and Condon wrote Oppenheimer forthright letters strongly condemning his action. Condon was the most critical of the three, having been "shocked beyond description."[41] Condon had known Oppenheimer since 1926, when they first became acquainted while living in the same *pension* as research students in Göttingen. He knew him "as a brilliant physicist and also as a person with a rather complicated personality."[42] In his letter, he charged Oppenheimer with involving other people in an effort to obtain immunity from harassment by HUAC for himself:

> I have lost a good deal of sleep trying to figure out how you could have talked this way about a man whom you have known so long, and of whom you know so well how good a physicist and good a citizen he is. One is tempted to feel that you are so foolish as to think you can buy immunity for yourself by turning informer. I hope this is not true. You know very well that once these people decide to go into your own dossier and make it public that it will make the "revelations" that you have made so far look pretty tame.

Condon urged him to write to the president of the University of Rochester to tell him that "Dr. Peters was all right." And should Peters lose his job, Condon insisted that Oppenheimer must "take him on at the Institute."[43]

Bethe's letter to Oppenheimer conveys his disappointment.[44] It reveals a perspicacious assessment of Oppenheimer's character and a clear-eyed judgment of security personnel and their methods of interrogation. But unlike Weisskopf, for Bethe the issue of Oppenheimer's status as a role model and charismatic leader was of secondary importance: it is what can be done to rectify the situation and help Peters that concerned him most. I quote the letter in its entirety:

> Dear Robert,
>   We are having a good conference here, with many new results and good discussions. But for many of us, the conference is overshadowed by our deep concern for the future of two of our colleagues, your brother Frank and Bernard Peters.
>   I can well understand how the damaging article about Peters in the Rochester paper came about. We all know the custom of the Un-American Committee to "leak" to the press parts of a wit-

ness's statement out of context, giving only the damaging evidence and leaving out the favorable. I can believe that your statement looked very different when it was given. Furthermore, I realize that your statement about Peters was made to de Silva in confidence, and that you had every reason to expect that it would never go beyond his files, and would certainly not come before a Congressional Committee. Moreover, your testimony was given in Executive Session, and there is no excuse for the Committee's publication of it. I think you have every reason for a sharp protest of this double (or multiple) gross indiscretion.

However, taking all this into account, I find it hard to understand that your testimony would even include such damning statements as were reported in the Rochester paper. I remember that you spoke in the most friendly terms to me about the Peters', and they certainly have considered you their friend. How could you represent his escape from Dachau as evidence for his inclination to "direct action" rather than a measure of self-defense against mortal peril? His approval of direct action (street fights) against the Nazis certainly does not imply approval of similar action against the democratic government of the United States; if anything, there exists a negative correlation between these attitudes. I probably know Peters much less well than you, but I have always found him quiet and reasonable. Ever since he came to Rochester, he has been working hard and successfully, and would hardly have any time for intensive political activity even if he were inclined towards it.

As a consequence of the article about your testimony, the University of Rochester has not yet renewed Peters' contract as an Assistant Professor. They have not dismissed him, so that the damage can still be repaired without attracting much public notice. You know that if he were actually dismissed, this would make it almost impossible, under present conditions, for him to find a job at another university.

Only you can help in this matter. If your actual testimony was as different from the newspaper report as I would like to believe, this should be easy. The first thing would obviously be to write a letter to President Valentine of Rochester, setting out your true opinion of Peters, and whether you think it would be dangerous if he were to continue to teach at Rochester. It would surely be helpful if a correcting statement from you could go to the Rochester paper, possibly an excerpt from your letter to Valentine. By doing so, you would not only help Peters who is a first class physicist, and yourself, but also our general fight against the custom of the Un-American Committee and other agencies to leak to the press the most damaging parts of the evidence against people who are substantially innocent. After this conference, I shall go to Los Alamos. Will you come there any time this summer? I hope you have a pleasant and cool and fruitful summer in Berkeley.

> With the best wishes
> yours
> Hans[45]

After the conference, Peters and Frank Oppenheimer went to see Oppenheimer in Berkeley. Their visit could not have been an easy one for Oppenheimer. At his trial in 1954 Oppenheimer gave his recollection of the general substance of his exchange with Peters. He recalled that Peters had asked him: "Was there a way in which I [Oppenheimer] could help him to keep his job at the University? He also said I had misunderstood him about his being a member of the Communist Party in Germany. He worked with the Communists, he was not ashamed of it, but he was not actually a member and nobody could prove that he was."[46] Peters also gave an account of his meeting with Oppenheimer in a letter to Weisskopf written a few weeks after the exchange between them had taken place:

> July 21, 1949.
> My talk with Robert was rather dismal. At first he refused to tell me whether the newspaper report was true or false. Then when I said there was nothing to discuss before he told me the truth he finally confirmed each statement printed there. He said it was a terrible mistake. He was not prepared for any questions. He had never before done anything as wrong. It did not seem to him that at the time that what he had said was bad, but that now seeing it in print, he could see that it was. When I asked why he concealed all this from me when I saw him in Princeton two days later and told me no questions were asked about me except one on AEC fellowships he got very red and said he had no explanation. He had called Rochester and found out that my position was not affected. So he said that he had fixed up everything and I had nothing more to worry about. He thought that he should not write to the Rochester papers. Just at that moment your letter arrived which impressed him and he changed his mind. The enclosed clipping is the result, in my opinion a not very successful piece of double-talk. He was obviously scared to tears of the hearings but this is hardly an explanation.
> Well I thank you for your letter you wrote him. I found it a rather sad experience to see a man whom I regarded very highly in such a state of moral despair.[47]

As a result of Peters's visit and of Bethe's, Condon's, and Weiss-kopf's intercession in his behalf, Oppenheimer wrote to the editor of the Rochester *Democrat and Chronicle* retracting some of the testimony he had given before HUAC. His letter was published on July 6. Oppenheimer declared that when he knew Peters as a graduate student at Berkeley he was "not only a brilliant student, but a man of strong moral principles and of high ethical standards. During those years his political views were radical. He expressed them freely, and sometimes without temperance." But, he added, "I have never known Dr. Peters to commit a dishonorable act, nor a disloyal one." He went on to say that Dr. Peters had recently informed him that he was right in believing that as a young man Peters "had participated in the Communist movement in Germany," but wrong in believing that he had ever been a member of the Communist party. He expressed "his profound regret that anything said [in confidential discussions between him and intelligence officers at Los Alamos in the context of wartime assignments] should have been so misconstrued, and so abused, that it could damage Dr. Peters and threaten his distinguished future career as a scientist." He concluded his letter with the following statement:

> Beyond this specific issue, there is ground for another, more general, and even greater concern. Political opinion, no matter how radical or how freely expressed, does not disqualify a scientist for a high career in science; it does not disqualify him as a teacher of science; it does not impugn his integrity nor his honor. We have seen in other countries criteria of political orthodoxy applied to ruin scientists, and to put an end to their work. This has brought with it the attrition of science. Even more, it has been part of the destruction of freedom of inquiry, and of political freedom itself. This is no path to follow for a people determined to stay free.[48]

Peters was right in characterizing the letter as "a not very successful piece of double-talk." Oppenheimer was being disingenuous. For to assert that "as a young man Peters had participated in the Communist movement in Germany" without specifying the context—particularly in the light of Bethe's reproach—and to characterize Peters's wartime political views regarding opposition to Nazi occupation as "radical" and expressed "sometimes without temperance" was to brand Peters a fellow traveler lacking prudence and self-control. Such a portrayal could only damage Peters politically, given the tenor of the times, and academically, when evaluating him as a teacher.

Five years later, during his questioning by Roger Robb before the Gray Board, Oppenheimer would equivocate about the intent and meaning of his letter to the Rochester *Democrat and Chronicle*. He there stated that his letter merely asserted that Peters had denied membership in the Communist party. It did not say that "I believe his denial."[49]

The University of Rochester, and its president Alan Valentine, did act honorably in the Peters case. After the disclosure of Oppenheimer's testimony, the university did not fire him, since, according to Valentine, there were no supported charges of "wrongful action on his part." In fact, shortly thereafter, the university "promoted him to a full professorship in the face of the House Committee's smears, and . . . this promotion came to him at an unusually young age because of his splendid research accomplishments."[50]

However, as a result of Oppenheimer's depositions to de Silva and his testimony before HUAC, Peters's freedom of movement was severely curtailed. In 1948, upon his arrival in France to attend a scientific conference, his passport was taken from him by an officer of the Sureté Générale, acting on orders from Washington, and Peters had to return to the United States without giving his lecture. He also experienced difficulties in having his Office of Naval Research (ONR) contract renewed; but it was renewed due to the outstanding quality of his work.[51] His inability to travel made it essentially impossible for him to carry out his research. At the time, he was investigating the amount of anti-matter present in the incoming cosmic radiation, and the research required performing experiments near the equator. He found himself unable to do so. In 1951 Peters accepted a professorship at the Tata Institute of Fundamental Research (TIFR) in Bombay. After the expiration of his American passport in 1952, he found himself once again unable to travel. "After it became apparent, that [he] could not elicit any reply to [his] letters to the State Department from Mr. Dulles on down (neither positive, nor negative, nor even dilatory) [he] resumed his previous nationality and returned his American citizenship." He did so with a "heavy heart."[52] His stay in India was characterized as "the most active scientific era in the history of TIFR." He there carried out important cosmic ray experiments and initiated a wide variety of cosmic ray geophysical studies.[53] In 1959 he became a permanent member of the Niels Bohr Institute in Copenhagen and continued research in high energy physics both there and at CERN in Switzerland. He was appointed director of the Danish Space Research Institute in 1968;[54] he there oversaw research on the Earth's magnetosphere and in cosmic rays. He held this post until his retirement in 1979. He died in 1993.[55]

The fact that Peters had a distinguished, productive, and influential career as a scientist in spite of Oppenheimer's testimony and his leaving Rochester is testimony to his impressive abilities, to his energies and his commitment to science. He probably would have had an even more prolific scientific career had he been able to stay in the United States. It certainly would have been an easier, but not necessarily a more rewarding, life.

Epilogue

The HUAC investigations of Bernard Peters, Joseph Weinberg, and David Bohm were to have further repercussions on Oppenheimer. In February 1950, on the heels of the disclosure of Klaus Fuchs's espionage, J. Edgar Hoover was asked to testify in front of the Joint Atomic Energy Committee. Fuchs's arrest in Great Britain had taken the American authorities by surprise. Hoover, in his appearance before the Joint Committee, spoke at length about Oppenheimer but made no connections between Fuchs's spying and Oppenheimer. The executive director of the Joint Committee at that time was William Borden. Hoover's testimony was Borden's first exposure to Oppenheimer's past political activities. Later that year, when reviewing the AEC's "most difficult" security cases, Borden came to see Oppenheimer's security file. When, in 1952, Weinberg's case—which was then going through the courts—threatened to expose Oppenheimer's past political affiliations and the reluctance on the part of HUAC and other congressional committees to investigate the matter and thus embarrass both the AEC and the Joint Committee, Borden launched a thorough investigation of Oppenheimer. In November 1952 he submitted his report to the chairman of the Joint Committee summarizing the content of Oppenheimer's dossier. Subsequently, Borden became involved in an assessment of the hydrogen bomb program and became convinced that Oppenheimer was responsible for delaying the project. The information Borden had acquired in preparing his 1952 report on Oppenheimer became the basis of his November 7, 1953, letter to J. Edgar Hoover, which initiated the steps leading to his denial to access classified documents and his "trial,"—a letter in which Borden claimed that "more probably than not" J. Robert Oppenheimer is an agent of the Soviet Union.[56]

THE PHILIP MORRISON CASE

On April 5, 1951, Cornell's acting president, T. P. Wright,[57] asked associate professor of physics Philip Morrison to come to his office because he wanted to apprize him "of the numerous letters [he] had received from trustees, alumni and others which have convinced [him] that [Morrison's off-campus political] activities are bringing great harm to Cornell."[58] Wright took the occasion to read to Morrison some of the sections in the 1940 Statement of Principles of the American Association of University Professors (AAUP) dealing with academic freedom. In particular, Wright laid great emphasis on the following passage:

> The college or university professor is a citizen, a member of a learned profession, and an officer of an educational institution. When he speaks as a citizen, he should be free from institutional censorship or discipline, but his special position in the community imposes special obligations. As a man of learning and an educational officer, he should remember *that the public may judge his profession and his institution by his utterances.* Hence he should at all times be accurate, should exercise appropriate restraint, should show respect for the opinions of others, and should make every effort to *indicate that he is not an institutional spokesman.*[59]

Taking his cue from another AAUP statement that coupled academic responsibility with academic freedom, Wright tried to convince Morrison that he should alter the character of his extracurricular political activities. Before the meeting, Wright had in fact drafted certain "suggestions," which he read to Morrison:

> 1. I urge you favorably to consider refraining from appearing on platforms in a sympathetic role with avowed or proved communists, because such action injures Cornell.
>
> 2. I urge and request you specifically to disassociate yourself from Cornell when expressing views in a controversial area outside your professional field.

His third recommendation was more specific. Wright asked Morrison "to refuse accepting sponsorship of any Cornell student group that may propose to organize for support of the Peace Crusade."[60] After he finished reading his "suggestions" to Morrison, Wright voiced the hope that Morrison would abide by them.

As recorded in the letter Wright sent to Morrison following their talk, Morrison had agreed with Wright at the meeting on the need "for associating academic freedom with academic responsibility." He also promised Wright that he would send him a copy of the principles of the Peace Crusade and copies of letters in which he had expressed his views on matters relevant to the discussion between them.

The meeting between Wright and Morrison was the first of a series of encounters between Morrison and the administration of Cornell University that would take place over the next seven years. Their goal was to try to find a modus vivendi that would accommodate Morrison's civil liberties, Cornell's commitment to academic freedom, and the political realities of the McCarthy era. Even though Morrison's background in physics and politics was similar to that of David Bohm, Frank Oppenheimer, Bernard Peters, and Joseph Weinberg, his experiences at Cornell turned out to be very different from theirs. But in this case, his public political activities became constrained and sharply limited. Also, a faculty committee was set up to investigate whether dismissal procedures should be initiated, and his promotion to full professorship was delayed. However, these steps were taken primarily to placate an influential minority on the Board of Trustees and some vocal alumni, as well as mitigating the criticism of the right-wing press. Morrison kept his job. And even though his position by 1953 had become very precarious, he always felt that the administrative officers with whom he had dealt—in particular, Acting President Wright and President Malott—had been sympathetic and supportive.[61] This is testimony to the enduring "liberal" character of the university set by White and Ezra Cornell, and of the institution's commitment to academic freedom under difficult circumstances. It is also a statement of the character and integrity of the men who set the tone of that institution during the difficult McCarthy era—Ezra Day, P. T. Wright, Deane Malott, Hans Bethe, and especially Robert R. Wilson and Dale Corson.

Morrison had obtained his Ph.D. in theoretical physics in 1940 working with Oppenheimer at Berkeley.[62] He had taught for a year at San Francisco State College before accepting an instructorship at the University of Illinois in 1941. In January 1943 he joined the Manhattan Project and worked with Eugene Wigner at the Metallurgical Laboratory at the University of Chicago. In early 1944, he was transferred to Los Alamos and became a member of the team that was responsible for the assembly of the plutonium bomb. He personally carried the plutonium core for the Trinity explosion to the test site at Alamogordo. In the summer of 1945, he went to Tinian to help assemble the bombs dropped on Hiroshima and Nagasaki. Together with Robert Serber he was one of the first Americans to visit Hiroshima and Nagasaki after the cessation of hostilities and to report to his colleagues at Los Alamos the devastation that had been wreaked and the calamity that had befallen the population of the two cities. At Los Alamos he had come to the attention of Robert Bacher and Bethe, who recommended him for an appointment at Cornell. The letter of recommendation that R. C. Gibbs, the head of the physics department, sent to Cornelius W. de Kiewiet, the dean of the college of arts and sciences at Cornell, gave the following assessment of Morrison: "He is generally regarded as very brilliant, has a wide range of interests, and is very promising both as a teacher and as an independent investigator. . . . He had numerous offers elsewhere and Cornell University is being warmly congratulated upon securing his appointment."[63] Morrison had accepted an appointment to the Cornell faculty as an assistant professor in July 1946. He was promoted to associate professor with tenure in the spring of 1948.

Like many of his friends at Berkeley during the 1930s, Morrison had joined the Communist party not only because of its anti-fascist stand, but also because it was championing the cause of the unemployed, denouncing discrimination against blacks, and organizing the California farm workers. He had become a member in 1936 and had remained one until 1942, at which time he severed his ties with the party. But in contradistinction to Bohm, Peters, Frank Oppenheimer, and other "atomic" scientists who had been affiliated with left-wing causes before and during the early stages of the war, Morrison remained politically involved after 1945. He was active in the efforts of the Federation of Atomic Scientists to have civilian control over atomic energy; he actively supported the Progressive party and the campaign of Henry Wallace, its presidential candidate, in 1948; and he was active in the peace movement of the late 1940s and 1950s.

Yet despite Morrison's prominence in postwar "left-wing" causes, neither HUAC nor the Senate Internal Security Subcommittee asked him to testify until 1953. He was, of course, fully aware of the tenor of the times. He had accepted a position in Ithaca rather than Berkeley or a university located in a large city because he knew Cornell's liberal tradition, he admired Bethe and Bacher, and—equally important—he felt that in this relatively isolated town in the rural, western part of New York State, he would be out of the national limelight.[64] He was careful not to bring about situations that would cause difficulties. Thus, he did not apply for a passport to travel overseas, nor did he maintain any connections with Los Alamos or engage in any activities that would have required a security clearance.[65] Yet, taking all this into account, it is not entirely clear why in fact Morrison was not asked to testify before HUAC. It is true that he never had any connections with the Radiation Laboratory, in Berkeley, which was the focus of HUAC's investigations. But his close association with Oppenheimer before and during the war surely made him a sensitive link for getting at Oppenheimer.

The fact that by 1948 Morrison had tenure made his case different from that of Bohm, Frank Oppenheimer, Peters, and Weinberg. According to AAUP guidelines, the dismissal of a tenured professor requires formal charges and due process. The only grounds for Cornell to initiate such a move would have had to be political, because Morrison's professional activities in physics and biophysics were outstanding. Ezra Day in 1948 had openly expressed himself against the retention on the Cornell faculty of members of the Communist party, either proved or avowed, since "the faculty should be composed of free, honest, competent, inquiring minds undertaking to find and disseminate the truth. No mind that is fettered or enslaved can possibly meet the requirements; hence, it seems to me to follow inevitably that anyone who admits allegiance to the communist party does not belong on a faculty such as ours."[66] But Cornell's administration was pretty sure that Morrison was not a Communist. The initiation of a dismissal procedure against him for his political views would have jeopardized Cornell's standing as one of the bastions of academic freedom and put it in an untenable position in academic circles. It would also have violated its charter and the tenets that had led to its founding.

Though not persecuted by congressional committees until 1953, Morrison was not immune from harassment by "the nation's professional anti-Communists, the right-wing journalists and congressional investigators who specialized in exposing [alleged] Communists and their followers."[67] They had intensified their efforts in the wake of the Korean conflict, particularly after President Truman had issued a proclamation on December 17, 1950, placing the country on a national emergency basis.[68] In early spring of 1951, an article by Kirkpatrick appeared in *Counterattack*[69] alleging the presence at Cornell of numerous Communists among the student body and on its faculty, and he pointed to Philip Morrison as one. The article prompted several members of Cornell's Board of Trustees to write to Wright, the acting president.[70] Two trustees, Victor Emanuel and Harold Bache, suggested that the faculty members mentioned in the article be dismissed. Both the chairman of the Board, Arthur Dean,[71] and Wright answered them. In his letter, Wright pointed out that Cornell abides by the AAUP guidelines and that it would be "exceptionally difficult" to discharge tenured members of the faculty unless one had actual evidence that they were members of the Communist party or were Communists. And even then it was not clear that this would constitute adequate grounds since neither the Congress of the United States nor the legislature of the state of New York has outlawed the Communist party as such. Furthermore, since all who knew Morrison "were unanimous in their belief that he was not a Communist," the case for dismissal against him would have to be based "on the assumption of guilt by association," and this was unacceptable to Wright.[72]

In his letter to Bache, Wright explained that Morrison had been characterized "as a pacifist and an almost fanatical advocate of peace." He added, "Possibly associated with this view is a certain guilt complex which many scientists possess because of leading parts that they may have played in the development of the atomic bomb." Since Wright had concluded that Morrison was not a Communist, and because he believed in the need for academic freedom "even for those who hold views to which we are violently opposed" the only approach open was "to attempt by persuasion to convince Professor Morrison that he should desist from taking part with persons who are admitted communists in activities which associates Cornell with mention of his name."[73]

The pressure on Wright increased after April 1, 1951, in the aftermath of the release to the national press of a 170-page *Report on the Communist "Peace" Offensive* by HUAC, several pages of which were devoted to Morrison's political activities. As a result of the publicity that the HUAC report elicited and of the extensive press coverage of Morrison's political actions following it and the *Counterattack* article, Arthur Dean wrote Wright to express his own anxiety and that of other alumni and board members. Dean's letter was one of the reasons why Wright had called Morrison in on April 5.

On April 17, Wright met with Bethe to convey to him his concern over Morrison's political activities and to obtain his counsel.[74] He again stressed that the "great number of letters" that he had received from trustees, alumni, and others were proof "that considerable harm was being done to the University" as a consequence of "the new light in which Cornell is placed as a result of Prof. Morrison's activities." Bethe told Wright that Morrison is "a passionate pacifist" and is reluctant to change his activities because of his sincere belief that he can contribute toward the cause of peace through them. He admitted that on several occasions "he had been annoyed by the unreasonableness of Professor Morrison's sponsorship of courses of action favored by the Soviet Union." In particular, Bethe had been bothered by Morrison's "charitable attitude towards Russia in the International Atomic Energy Control Debate where he expressed the opinion that Russia would come around to the Baruch Plan eventually," and that when they did not do so, "he [Morrison] changed his tune somewhat towards sponsoring the alternative Russian plan."[75] Bethe confirmed and attached great importance to the fact that Morrison had approved of the entrance of the United Nations into the Korean Policies action, but indicated that Morrison was now supporting the Peace Crusade, a crusade that, according to Wright, would end the war short of its objectives.

Wright then asked Bethe to read the letter he had written to Morrison on April 5. Bethe found it "reasonable" and made the "tentative suggestion, . . . that it might be appropriate to demand that Morrison clear in advance statements that he was intending to make at extra-curricular activities outside his field." And upon rereading the Statement of Principles of the AAUP, Bethe felt that Cornell "should demand appropriate restraint from Morrison." Furthermore, since Morrison had as yet not sent his reply to Wright and Bethe was of the opinion that "certainly Morrison should answer Wright's letter," he indicated that he would talk to him. Wright then gave Bethe a copy of his letter to Morrison and asked him to discuss it, "along with the whole matter" with Wilson.[76]

Wright, in fact, had already written to Wilson asking him, as director of the Newman Laboratory, to evaluate Morrison's activities in the physics department and in the Nuclear Lab. After reading Wright's April 5 letter to Morrison, which Bethe had showed him, Wilson answered Wright's request on April 19. Commenting on Morrison's teaching, Wilson stated that "as a teacher Morrison certainly competes with Professor Bethe for first honors in our department. His lectures are models of clarity and he uses an elegance of speech which goes far to hold the listener's interest to difficult subjects." He characterized Morrison's presentations at colloquia and seminars "real intellectual occasions." With a tinge of irony, he noted that because Morrison is "a very tender-hearted fellow" and finds it difficult to refuse being someone's adviser, he perhaps had too many graduate students doing theses with him—far more than any other faculty member. Some of the worst students in the department were among them, and the latter "impose more of a drain on his time than the good men who are also attracted to him." But he stressed that "without Morrison our other theoretical physicists, such as Bethe for example, would be completely snowed under directing graduate work, and in that case Cornell would become an unattractive place." Yet, despite his "crushing teaching" load, Morrison still found time to contribute in a major way to original research. In Wilson's estimation, all this added up to "a very outstanding" faculty member and everyone in the Department of Physics and in the Laboratory of Nuclear Studies was "very proud to have him as a colleague."

Wilson concluded his letter by addressing the issues that Wright had raised in his letter to Morrison and implicitly exhorted him not to be fainthearted in standing up for the academic freedom of faculty members:

It is true that [Morrison] has felt impelled to participate in somewhat dubious political undertakings and I am amazed that these political undertakings do not interfere with his normal duties. Perhaps he would be somewhat more creative if he did not put so much intensity into politics. However, the politics is done during a very small fraction of his spare time, and I have always considered it none of my business.

Morrison has an extremely liberal point of view and I have never found him anything but perfectly rational and certainly he has never injected anything that could be even construed slightly as subversive in his political discussions in the Laboratory or in social functions related to the Laboratory. It is my strong conviction that he is sincerely working for peace in the best way that he knows how, even though many of us do not agree with his methods.

After his meeting with Wright, Bethe had discussed the issues involved with Morrison, who in turn sat down and wrote a long reply to Wright "to place in permanent form" some of the ideas that had entered their exchange on April 5. He also thanked him for "the reasoned and understanding nature of our discussion of these difficult and controversial matters."[77] In his letter, Morrison declared that he would do his best to act in conformity with the spirit of the three suggestions Wright had made.[78] He accepted as "unexceptionable" the passages that Wright had quoted from the 1940 Statement of Principles of the AAUP and assured Wright that he had always tried to fulfill these principles "without qualification or reservation." If he had erred in any way in not complying with the AAUP guidelines, he "would indeed be pleased to take any steps possible to correct such errors, and to act in such a way as to prevent their repetition." He pointed out that thus far no specific instances had been brought to his attention.[79] But, he added,

The problem is this: out of my whole experience in life, and especially out of the events which culminated in my walking through the rubble of Hiroshima, I have gained the deep conviction that in the true interests of America, my country, it is urgent that some voices speak for peace, even in times of crisis and even in the face of bitter opposition. The catastrophe of Hi-

roshima, matchless in human misery and in the profound
moral erosion of a world perverting such powers, can come to
the United States. It is not easy to take such a stand, particularly
in a world where great-power conflict is the way of international
life, without angering many who see in the insistence upon
peace a surrender of national interest. But I am convinced that
the only real security for America is peace, and the best patriots
are those who urge a policy of peaceful settlement. If the people
once recognize the real meaning of another world war, the diplo-
mats are clever enough to forge a peace on mutually acceptable
terms. I do not believe that the Cornell tradition would have me
surrender my convictions on such grave issues, upon the resolution of which not only the welfare of the University but the
safety of our whole country depends. I know moreover that not
all patriotic Americans disagree with my views, though as yet
they are shared by only a minority. This is both inescapable and
proper; in a democracy ideas may begin with a few, but spread
to a majority in time. And was it not our own Carl Becker who
defined a professor as "a man who thinks otherwise"?

Wright promptly answered Morrison on the 23rd of April to express
his appreciation for the care with which Morrison had set down his
views regarding the suggestions Wright had made. He also voiced his
delight by noting "the very wide area of agreement that there now
appears to be between us." Wright went on to say that "broadly stated,
it would seem that we are in complete agreement on fundamentals."
The disagreement concerned Wright's differing assessment of the activities of the American Peace Crusade, of which Morrison was one of
the sponsors. Wright believed "that those of intellect, intelligence and
integrity who have associated themselves with it have to a considerable extent 'been used' by others who are much more interested in
promoting the aims of Soviet Russia than aims of the Peace Crusade
itself." Since Morrison's association with "the American Peace Crusade and the nature of the publicity attendant to its meetings,
whether rightly or wrongly, is working substantial injury to Cornell
University, and . . . to [Morrison] himself," Wright again urged him to
"refrain from further activities connected with the American Peace
Crusade." But he reemphasized the fact that what he was giving is "a
recommendation, and a course of action that he strongly urged," but
that "the basic principles of academic freedom, with which we are in
agreement, impose the final responsibility on you, the professor."[80]

Morrison heeded Wright's recommendations. Yet he remained active in the civil rights and peace movements, very conscientiously

and judiciously trying to comply with the spirit of the agreement. However, given the temper of the times—the period from 1952 to 1954 was the nadir of the McCarthy era—it was likely that he would come under attack again. This indeed happened. On December 29, 1952, J. B. Matthews, one of the most rabid of the "professional" anti-Communists, identified him to the Senate Internal Security Subcommittee as someone who was "currently active in this entire-Communist front movement." Matthews's charge was widely reported in the national press the next day, and once again Morrison's "extracurricular activities" became a matter of concern. Cornell's new president, Deane W. Malott, asked Morrison to meet with him on December 30, and they agreed that it would be helpful if a special faculty committee were appointed to "consider problems arising from unfavorable publicity received by Professor Morrison, primarily because of the testimony of J. B. Matthews before the Senate Internal Security Subcommittee." Such a committee was appointed by Malott,[81] and it was asked to consider whether "formal charges under the dismissal procedures of the By-Laws should be instigated."[82] The committee reviewed the "extensive files on this matter accumulated by the University Administration" and met twice with Morrison in "long sessions" in January 1953. It issued a report at the end of that month stating that Morrison had met with the committee "willingly and cooperatively" and had impressed its members "as being straightforward, helpful, and sincere."[83] The report recorded that Morrison had made the following statements:[84]

1. That he believed that he had carried out the spirit of Dr. Wright's requests made in 1951.

2. That he was one of the organizers of the American Peace Crusade and that he has continued correspondence with and for that organization but has not attended its meetings since the one in Chicago in July 1951. . . .

3. . . . that he was not anxious to bring about a change in the form of our government by any means let alone violent ones. . . .

4. Asked forthrightly if he were a Communist, he said, "I certainly am not—but neither am I wholly opposed to them." (He left the inference that he had much stronger sympathies with the Communist movement in his student days in Pittsburgh during the Depression.)
    When asked for evidence that he is not a Communist, he cited (1) his "opposition to Vishinski's stand . . . for a Korean

peace" and (2) that in his biophysical work, an area in which he has professional competence, he is definitely not a supporter of the Lysenko theory of genetics—a party-line concept.

5. Asked if he would be loyal if war were declared between the U.S. and the USSR, he said that he would certainly oppose a declaration of war but would support the U.S. if war were declared.

6. He is, and has been for years, a member of the American Labor Party, a legal political party.

The questioning by the committee had been designed to get on the record Morrison's statements that the administration could use to defend him, and itself, from attacks by trustees, alumni, and the press. In its concluding statement, the committee characterized Morrison as "an objective scientist, pursuing scientific truth with an open mind." It added patronizingly:

He has a strong feeling of guilt for his work on the atomic bomb and the loss of life and destruction it wrought upon Hiroshima. This sense of guilt and horror may easily have left a blind or weakened spot in his capacity to think—one which affects his social and political concepts while having no effect on the scientific work at which he is indisputably brilliant. At any rate he has become a rabid seeker of peace, even to the point of willingness to make concessions which a majority of his fellow citizens would not make. . . . At present he appears to be in a state of considerable doubt as to whether his peace efforts are bearing any fruit, and whether his methods have been effective.

The committee affirmed that Morrison was "not subversive" and that "his scientific work and his more ingenuous extra-curricular activities have not been blended." Since neither publicly nor in his contacts with students has he advocated "the overthrow of the government of the United States by force and violence, or the accomplishment of political change by a means not permitted by the Constitution of the United States or the State of New York," he should "not be charged with any activities which would make him guilty of such misfeasance or malfeasance as make him unfit to participate in the relationship of teacher to student." They judged that "his apparently honest motives are not censurable, but [that] his methods and associations have left him open to severe censure"; and that "only by exercising extreme discretion will he be able to avoid news comments adverse alike to himself and to Cornell."[85] Reflecting the intensification of the anti-Communist hysteria in the aftermath of the Korean war, Morrison was being put on a more stringent probation. At this very same time, Senator William Jenner's Senate Internal Security Subcommittee (SISS) was embarking on its investigation of educators, and Cornell feared that some of its faculty members might be asked to testify.

In February 1953, *Counterattack* published another article on Morrison pointing to his continued activities in the peace movement. Jenner's SISS was then just preparing to hold hearings in the Boston area. During the spring term of 1953, Morrison was on sabbatical leave as a visiting professor at MIT, and it was expected, in view of the *Counterattack* article, that Jenner would ask him to appear before the subcommittee. Both Cornell's administration and Morrison braced themselves for the repercussions of Morrison's being asked to testify. In March, Cornell's Policy Committee voted to set up a special, essentially secret, Subcommittee on Academic Problems Arising from Governmental Investigations (SAPAGI). For his part, Morrison retained Arthur Sutherland, a conservative Republican, as his lawyer when Jenner asked him to appear before his subcommittee. Sutherland, who had just left Cornell Law School in Ithaca to accept an appointment at Harvard Law School, was not prepared for the way the Committee operated. He became deeply disturbed by the tactics used to obtain some of its information—such as opening Morrison's mail and tapping his phone without a warrant[86]—and by the unsubstantiated accusations it was willing to make. Though he disagreed with many of the political statements that Morrison had made in the past, Sutherland passionately believed in his right to advocate them. Despite Sutherland's recommendation against it, Morrison elected to plead the "diminished fifth," that is, talk freely about himself but not about others. Morrison had decided to do so because he was convinced that he would lose his job if he were to invoke the Fifth Amendment as his defense. In a closed, preliminary meeting, Jenner asked Morrison numerous questions about Oppenheimer, which he refused to answer. But in a second, public session, Jenner refrained from asking any questions about Oppenheimer.[87] After that hearing, Sutherland wrote John D. MacDonald, the chairman of Cornell's secret subcommittee, to inform him that Morrison "answered every question that was asked, testified simply and courteously, and altogether made a good impression." He concluded by saying, "I see nothing which should suggest to Cornell any reason to make trouble for Philip."[88] The letter did allay Cornell's fears temporarily. New concerns arose when, in July, SISS, in a report on its activities, devoted an entire section to Morrison and charged, among other things, that on the basis of its "limited access" to his security questionnaire it has "learned that he had withheld his Communist Party membership from the security authorities[20]" of the Manhattan Project.[89] Footnote 20 in the report asserted that when asked to list all the organizations and to state their nature (such as athletic, fraternal, labor, political, professional) of which he was or had been a member since 1930, "Professor Morrison listed only American Federation of Teachers and Pi Mu Epsilon." And when asked under question 6 of the questionnaire, "Are you or have you ever been a member of any political party or organization which now advocates or has ever advocated the overthrow of the constitutional government of the United States?," Professor Morrison's answer was "No." The source of this information was claimed to be a letter dated June 30, 1953, from Gordon Dean, the chairman of the AEC to William E. Jenner, the chairman of SISS.

MacDonald's subcommittee took notice of the charge and asked Bethe to look into the matter. On a trip to Los Alamos in the fall of 1953, Bethe was able to look at Morrison's security questionnaires at Los Alamos and sent MacDonald not only a statement of the relevant answers that Morrison had given, but also the exact formulation of the questions. "These were directly copied by me from the originals at Los Alamos." Bethe wrote MacDonald:

> Please note the divergence between statement A, which is all in the present, and the second paragraph of footnote 20 of the Jenner Committee report, which alleges Philip Morrison's statement referred to the past as well as to the present. A statement of the kind quoted by the Jenner Committee's footnote was introduced into the questionnaires only in 1949—see part D of the enclosed document.
>
> The first part of the Jenner's Committee's footnote 20 is in reasonable agreement with the actual statement by Morrison under heading B of the enclosed list.
>
> The information on the enclosed sheet is of course confidential.[90]

Bethe was thus able to dispel the committee's apprehension by demonstrating that SISS had based its conclusions on what it took to be Morrison's answers to a questionnaire that was not distributed until several years *after* Morrison had left Los Alamos.[91]

But Morrison's difficulties continued to mount during the autumn. On April 22, 1953, U.S. Attorney General Herbert Brownell, Jr., had announced that twelve new organizations had been added to the list of organizations that had been declared subversive. Several alumni were quick to point out to Malott that Morrison was a member of three of them. During the summer, *Counterattack* renewed its onslaught on Morrison and chastised Malott—whom it quoted as having "committed Cornell to the position that neither Communist Party members nor fellow travelers have any right to teach at an American University"—for keeping Morrison at Cornell.[92] After Morrison addressed a National Council of the Arts, Sciences and Professions meeting on September 27 at Carnegie Hall[93] and proposed a resolution asking the United States and the Soviet Union to adopt "a spirit of understanding and conciliation in order to solve the problem of international regulation of atomic weapons," Malott asked him for an explanation of this violation of his agreement to exercise extreme discretion. Before Morrison could answer, *Counterattack* again stridently berated Morrison and Cornell. This was the straw that broke the camel's back. On December 3, 1953, Malott wrote Morrison:

> The continued embarrassment which you have caused Cornell University through the years has led me most regretfully to the conclusion that some action must be taken to protect the good name of the institution and those who are members of it from the continuing concern for its integrity caused by your repeated backing and support of allegedly subversive organizations. . . .
>
> You have appeared from time to time to indicate a willingness to be cooperative which promise has repeatedly failed of fulfillment.
>
> . . . I am therefore asking you to show cause in writing to me why I should not institute proceedings for your dismissal from the University.[94]

Morrison was stunned. And so was Bethe. He went to see Malott to express his dismay and asked Herrell DeGraff, who had become the chairman of SAPAGI, for an invitation to appear before it. After his meeting with Bethe, Malott wrote DeGraff in a letter marked "Confidential":

> Professor Bethe seemed very much agitated and talked all over the lot to me because of the hard-boiled attitude I have taken, and he seems to be somewhat unhappy about it. On the other hand, it seems difficult for him to focus himself, I think, specifically in a problem in human relations. He is too much the physical scientist.
>
> I rather hope you will not have Professor Bethe at your meeting; although, if you think it wise, I would certainly have no objection.[95]

Bethe was not the only one upset by Malott's course of action in the physics department. In his letter to Morrison, Malott had indicated that he had consulted with SAPAGI and that the recommendation had originated with them. Morrison had shown Malott's letter to his colleague Dale Corson and had discussed its content with him. Corson then wrote Malott urging him to withhold the formal charges because he was convinced that if they were to be brought against Morrison, "the University will suffer irreparable damage. . . . The action . . . could result only in a bitter struggle among the faculty, administration and friends of the University" with much loss and little gain. Morrison had indicated to Corson his willingness to cooperate and to abide by whatever steps Malott considered necessary to resolve the issue. Corson hoped that these steps would be clearly defined and would not "require complete abridgement of [Morrison's] right of free speech. I believe as citizens we all have a right and a duty to make our position known on public issues." Corson raised two other points concerning the procedures in effect in Morrison's case. He disapproved of the "informal" basis on which the committee was established, its "informal" use by Malott, and the fact that no written records of its proceedings had been provided. In addition, he deplored the failure of any member of the committee to speak to Morrison personally "to find what he considered the facts to be in the matters under consideration." Surely, "one who is the object of charges so severe has the right to be heard personally before the charges are made."[96]

Morrison wrote to Malott on December 16 explaining his activities and offered to "curtail sharply [his] associations with those organizations whose public standing has been impaired by the legal action of the Attorney General." He proposed that Malott and he draw up such a list, and that he would then promise not to join them, speak at their meetings, or let his name be used by them.[97] Malott took over a month to answer him, in order to consult with both members of the faculty and the Board of Trustees. The tone of his reply was much friendlier:

> I was very happy to receive [your letter] to which I have given careful and prolonged attention.
>
> I have not only studied your letter with great interest, but appreciate very much the apparently frank statement you have made. The most important part of your letter, the proposals which you make beginning on page four, are particularly interesting.

Malott counseled Morrison that he not only had to "curtail sharply" his associations with organizations on the attorney general's list but that his present activities had to be "especially discreet, to protect adequately not only the University as a whole, but your colleagues and your own professional standing." But Malott was unwilling to draw up any list of organizations which Morrison should avoid, for it would be inappropriate for an administrative officer to exercise such control over any member of the faculty. Moreover, Malott told Morrison "you must be, and I am sure in the last analysis would wish to be, the sole judge in the matter of your actions; it is thus my inescapable responsibility to decide whether you are transgressing the limits of tolerance which the University should extend to you, in view of all the circumstances." But in effect, and essentially explicitly, Malott was advising Morrison not to have *any* association with organizations lying outside his professional field. And having Morrison's compliance may well have been the intent of the "hard-boiled" attitude he had taken in threatening dismissal procedures. Malott concluded on a sanguine note:

> I do want to say again that I am deeply grateful for your letter, for its careful analysis with much of which I am in agreement.
> Also, I am deeply appreciative of your distinguished service as a scholar and a teacher in the field of physics. I want very much to have a growing confidence in your attitude and in your own spirit of cooperation, to the end that we may avoid future embarrassment.[98]

Morrison was again on probation. He essentially totally curtailed his public political activities and the attacks on him in the right-wing press ceased. However, some of this also reflected the relaxation of the witch hunting of Communists and fellow travelers following McCarthy's downfall after the Army-McCarthy hearings in April–June 1954.

But Morrison's difficulties at Cornell did not cease. In 1954 the Department of Physics recommended his promotion to a full professorship, but Malott refused to transmit the recommendation to the Board of trustees, knowing that the promotion would meet stiff opposition. The following year the department again requested his promotion, and again Malott demurred. But the department felt so strongly about the matter that it informed Malott that it would not recommend any other promotion to that rank "until favorable action is taken in Professor Morrison's case." This forced Malott's hand, and on April 21, 1956, he submitted his recommendation of Morrison's promotion to the Board.[99] The executive committee of the Board of Trustees then requested a full review of Morrison's past political activities. A confidential "Memorandum re Professor X" was drawn up by William Littlewood, whom John L. Collyer, the chairman of the Board of Trustees, had appointed as chairman of the ad hoc committee "to review the case of Professor Morrison with reference to any improper association with subversive activities."[100] In October 1956, Morrison was questioned under oath for two days by the ad hoc committee, and once again he was asked about all the supposedly subversive political activities Morrison had engaged in since 1946. For all their differences in political viewpoints and commitments, Morrison was able to convince the members of the ad hoc committee that he was a loyal citizen. But both Morrison and Corson, who was now chairman of the physics department and had accompanied him to the meetings, were upset about the adversarial nature of the inquiry. As Corson told the Trustees, "the fundamental issue is the place where we draw the line in requiring the conformity of a faculty member in areas outside that of his primary professional competence. . . . My feeling is that the line has been drawn rather far on the wrong side."[101]

The entire proceedings may well have been a staged affair to assuage a small, staunchly conservative minority on the board. The recommendation for promotion was approved by the board when the ad hoc committee "cleared" Morrison. And thus ended the inquisition of one of the most courageous defenders of civil liberties and one of the most forceful and outspoken advocates of a peaceful solution to the Cold War during the McCarthy era.

SOME CONCLUDING COMMENTS

The contrast between Oppenheimer and Bethe in the Peters and Morrison cases, respectively, is striking but a comparison is somewhat unfair. Any judicious assessment of their comportment must recognize the harshness and brutality of the anti-Communist crusade that was being waged; HUAC, McCarthy, and Jenner were only the most visible components.[102] It must take into account Oppenheimer's vulnerability by virtue of his "leftist" views and political activities during the 1930s—views that reflected his dread of Hitler and the National Socialist state and his fear that fascism would spread—his vulnerability by virtue of his brother's and his wife's past membership in the Communist party, and his enduring love and friendship for his brother. Oppenheimer's prewar involvement with the politics of the Popular Front made him fated to suffer for virtues that in the postwar context came to be judged as errors. But his stand in the late 1930s was similar to that of many other concerned highminded individuals of his generation. Oppenheimer's engagement is to be contrasted with Bethe's lack of political involvement during the 1930s because he was not a citizen of the United States. A judicious evaluation must take into account the vicious campaign that was being waged against Oppenheimer in order to eliminate his influence within the government; his passionate efforts to moderate the nuclear arms race; and his status as an influential *insider* as compared to Bethe's position as an *outsider* in the corridors of power. Any statement that Oppenheimer might have made in support of Morrison would probably have only made Morrison more vulnerable. Thus Edward Teller told the FBI in July 1949 that

> Morrison is an extremely gifted and brilliant man who is reliable, congenial and a brilliant speaker . . . [who] made very valuable contributions to the [atomic bomb] project. [He, Teller] was unable to vouch for Morrison's loyalty and stated that Morrison has the reputation among physicists of being extremely far to the left. [Teller] stated that Oppenheimer, Robert Serber and Morrison are considered the three most extreme leftists among physicists. [Teller] stated that most of Oppenheimer's students at Berkeley had absorbed Oppenheimer's leftist views.[103]

The mounting pressure on both Oppenheimer and on Morrison from 1950 on reflected the fact that McCarthy had become a major political force. In the spring of 1950, an FBI informant, Sylvia Crouch, had accused Oppenheimer of being a Communist party member in the late 1930s, and the story received ample coverage in the national press.[104] By the fall of 1952, the attacks on Oppenheimer had become frequent and public.[105] A widely publicized article in *Fortune* in May 1953 accused him of being responsible for an eighteen-month delay in the development of the hydrogen bomb.[106] All these issues have been dealt with at length in the literature.[107] I shall only comment on two aspects of the difference in the comportment of Bethe and Oppenheimer.

I would like to suggest that one of the elements, a crucial one, that made it possible for Bethe to act the way he did was the self-confidence he had acquired from his performance as a student in the Gymnasium and in Sommerfeld's seminar, and also Sommerfeld's encouragement and strong recommendations. This self-confidence had extended to all facets of his life, and gave him the courage and fortitude to act resolutely and forthrightly in political and moral matters. Oppenheimer never received this kind of approbation from his teachers—neither from Kemble nor from Bridgman, his mentors at Harvard, nor later from J. J. Thomson, Born, Pauli, or Ehrenfest.

Another important factor was the community of which Bethe was a part at Cornell. One should not underestimate the importance of the resonance of Bethe's views with those of his close associates in the Newman Lab (in particular Robert R. Wilson and Dale Corson), with those of most of his colleagues in the Department of Physics, and with those of many of the tenured faculty at Cornell. Bethe was a member of a community that was conscious of its responsibility in upholding the tenets of academic freedom upon which Cornell had been founded, a community whose members rallied to the support of one another both at the collective and at the personal level. I do not believe that this was the case for Oppenheimer at the Institute for Advanced Study. There was probably no one to whom Oppenheimer could turn, if such had been his inclination, to discuss the difficult issues he was confronting.[108] This is a reflection of Oppenheimer's personality, not of the political views of the faculty at the Institute. But it does indicate that a chasm existed at the social level that intensified Oppenheimer's isolation. It is to this theme of the individual and the community—a subject frequently addressed by Oppenheimer in his lectures after World War II[109]—that I will return in the Epilogue.

5.

NUCLEAR WEAPONS

> Everyone who is aware of the possibility of mankind's
> self-destruction must resist that possibility to the
> utmost.... The resistance against the self-destructive
> consequences of man's technical control of nature
> must come through acts which unite religious, moral,
> and political concerns, and which are performed in
> imaginative wisdom and courage.
>
> —Paul Tillich

ATOMIC BOMBS

How to act morally has been one of Bethe's constant concerns. The
boundary between the moral and the political have never been sharp
for him. He learned early that Kant's individualistic categorical imperative, though morally binding, was often not effective in halting
or eradicating evils in the world. The futility of James Franck's resignation as director of the Göttingen institute for experimental physics—in contrast to the efficacy of the collective efforts of Niels Bohr,
Ernest Rutherford, and their colleagues in Europe and of Livingston
Farrand[1] and others in the United States in finding suitable positions
for the members of the professorate who had been dismissed from
their posts with the advent of Hitler—left its mark on Bethe.

If Hitler sensitized Bethe to the need for collective action, it was
Los Alamos and the dangers that nuclear weapons posed to mankind
that have forced him constantly to be involved in moral and political
issues. Was it right to use an atomic bomb on Hiroshima? On Nagasaki? Should one build hydrogen bombs? In what way do these weapons of mass destruction—which, if used, would affect the future of
the planet in an irreversible manner—force us to rethink the waging
of war and the maintenance of peace? Should one build shelters for
the civilian population? Should one build anti-ballistic missiles? And
at the personal level, what is the proper social attitude of the scientist
and the engineer? "Should he just follow directions from society, from
government, from industrial management, and develop what he is
asked to do? [i.e., should he be] on tap but not on top? Or should he

feel the full responsibility for the social consequences of his inventions which would imply that the scientists themselves would have to decide which things are good for the world and which are not."[2] If the latter, should a scientist participate in the research necessary to develop nuclear weapons, missiles, and the like? Is there a difference in doing so in peacetime and in wartime?

The discussions to confront the moral and political dimensions of these issues always started at home. One of the striking differences between the refugee and European-born scientists who worked at Los Alamos and their American colleagues was that the married European-born scientists told their wives of the general nature of the project they were engaged in.[3] This was the rule for them, but the exception for married American scientists.[4] Bethe has always clarified key moral and political issues through intense discussions with his wife. At times—as in his decision to decline Edward Teller's urgent and insistent request to come to Los Alamos to work on the Super in the fall of 1949—her views were crucial.

On many occasions, Bethe has stressed that one must make a distinction between the development of the atom bomb during World War II and the subsequent work on the hydrogen bomb. He also stresses the distinction between working during World War II on radar research at the Radiation Laboratory and working on designing atomic bombs at Los Alamos. "Radar did not create a new weapon of destruction, but did make existing weapons more effective. . . . Radar was decisive during the entire European war and also the Japanese war. That was a case where the scientist should clearly participate." Whether to participate in the atomic bomb project was not so clear-cut. On the one hand, an atomic bomb was a completely new weapon with unprecedented mass destruction capabilities, one that "surely would forever after trouble us." On the other hand, the United States was engaged in a life-and-death struggle with Nazi Germany. It was clear that if Germany obtained nuclear weapons before Great Britain and the United States, and would use them to win the war, it would mean the Nazi and fascist conquest of the world and the end of Western civilization. The decision to work on the bomb was a difficult matter of conscience for Bethe, but he decided that he should join the project.

The concept of an atomic bomb and its feasibility became immediately apparent after the news of the fission of uranium by Hahn and Strassman was reported by Bohr to the Washington Conference on Theoretical Physics in January 1939. Bethe was in attendance, and although excited by the news, he did not think that the physics involved was of a fundamental nature. The possibility of a fission bomb became immediately apparent, but neither he, nor any other physicist at the conference, thought of raising the question of their ethical responsibility. From February 1939, until the end of August 1939 the international channels of communication between physicists were open, and publications related to nuclear fission were unrestricted. There then existed the possibility that, collectively, physicists might quietly declare a tacit moratorium on the construction of nuclear weapons. To bring this about required courage and moral leadership. But no one in the know and of sufficient stature to convince the community—neither Bohr, nor Fermi, nor Heisenberg, nor Kapitza, nor Rutherford—came forward. After the invasion of Poland by Hitler in September 1939 and the declaration of war by France and England, it was too late.[5] By then, Bohr and Wheeler had explicated the mechanism of fission, the possibility of a nuclear chain reaction had been verified experimentally, and the mistrust of the activities of German scientists was too great to be able to try to effect a moratorium.

Initially, Bethe was not convinced that it would be possible to build a nuclear weapon using the chain reaction properties of an assembly of U^{235} and affect the course of the war, for he believed the technological problems involved were too enormous to be undertaken at that time.[6] After the outbreak of hostilities, he felt that there were more urgent and important war-related problems to be addressed, in particular armor penetration and radar. Work on radar took him to the Radiation Lab at MIT in 1942. Although he was aware of the efforts by Fermi, Szilard, Wigner, and others to build a reactor at the University of Chicago, Bethe was not familiar with the developments there and the progress that had been achieved, nor of its implications for the timescale for building an atomic bomb.

The feasibility of a fast neutron U^{235} fission bomb was first analyzed by Rudolf Peierls and Otto Frisch and their associates in England. The MAUD Committee which was appointed by the British government on the heels of the memorandum Peierls and Frisch had written in March 1940, concluded that indeed a bomb was feasible. The MAUD report raised the critical mass to something in the neighborhood of three kilograms and estimated that the timescale for its construction would be two and a half years at a cost of roughly £5 million. The MAUD report was made available to the United States government in October 1941.[7] It persuaded Vannevar Bush and James Bryant Conant to proceed with dispatch and high priority on the uranium and plutonium bomb project. A few days before the Pearl Harbor attack the decision was taken by them to give Arthur Compton, Ernest Lawrence, and Harold Urey key responsibilities for the project.[8] During the summer of 1942, Bethe was invited by Oppenheimer to be part of the workshop concerned with the design of both a uranium and a plutonium bomb. Bethe accepted after making sure that there were other people at the Rad Lab who could do what he had been doing. The presence at the Rad Lab of Julian Schwinger—who, according to Bethe, "could do these things even better than I"—made this possible.

Bethe went to Berkeley from Boston via Chicago. In Chicago he was joined by Edward Teller, who since January 1942 had been working at the Met Lab. Teller had also been invited to the summer workshop by Oppenheimer. On the two-day train ride from Chicago to San Francisco, Teller informed Bethe of everything that had been happening on the atomic bomb project. Both Bethe and Teller were convinced that the Germans were working on a reactor and toward a uranium bomb, but neither one had any idea how far they had gotten. They knew at the time that the Germans were very interested in getting heavy water from Norway, and that this was an excellent moderator in making a nuclear reactor. The group at the Met Lab—Enrico Fermi, Les Szilard, Eugene Wigner—had decided not to use heavy water because they had discovered that extremely pure graphite was equally effective and much easier and cheaper to produce. The Germans, as was later found out, had to use heavy water, because they had used unpurified graphite in their neutron absorption measurement and had concluded that carbon could not be used as a moderator. What became clear to Bethe from his conversation with Teller was that a uranium weapon was not pure fantasy, but that it could be built during the war.

The summer workshop made a reliable determination of the amount of fissionable material required for a bomb and the estimation of its efficiency.[9] It also spent a fair amount of time investigating the feasibility of a "Super," that is, of a fusion bomb that would be ignited by a (yet nonexistent) fission bomb.[10] The results obtained by the Berkeley study group, and the clear indication during the summer of 1942 that the pile that Fermi was constructing under Stagg Field in Chicago would soon go critical, confirmed the feasibility of a uranium bomb. With the concurrence of his wife, in early January 1943 Bethe decided that he would join Oppenheimer at Los Alamos. In the spring of 1943, the laboratory was established to design, develop, and assemble an atomic bomb. Bethe arrived there in late March 1943. Soon thereafter, Oppenheimer asked him to head the theoretical physics division—one of the seven divisions of the laboratory.

During the war, there was never any question in Bethe's mind that the bomb had to be built, for otherwise "the Germans could one day drop atom bombs on us." Nuclear fission had been discovered in Germany, and there were very good reasons to assume that the Germans had been working on an atomic bomb since 1939. Given the caliber of the German scientists working on the problem—Heisenberg, von Weizsäcker, Bothe, Harteck—it was likely that they were much ahead of the Anglo-American effort which began in earnest only in 1941. Bethe, and all the scientists at Los Alamos, were convinced that they were in a desperate race, and that not only did the work have to be done, it had to done as expeditiously as possible. The doubts came only later: for a few of them, in the winter of 1944 when victory over Germany seemed evident;[11] for many of them, after Trinity, the code name for the bomb's test at Alamogordo in the New Mexico desert.

Until Trinity, most American physicists at Los Alamos were unaware that British intelligence had obtained fairly conclusive information that Germany had stopped working on an atomic bomb; nor were they informed of the findings of the Alsos Mission about the status of Germany's atomic bomb project.[12] Bethe was not apprized of the nonexistence of a German bomb project until just before VE-Day in May 1945. But in fact, by late fall 1944, as reports from Goudsmit and the Alsos Mission became available, it had become clear, though not conclusively so, to those privy to the information the mission had gleaned that Germany was not developing an atomic bomb, and that no large-scale industrial effort to manufacture either U^{235} or plutonium had been attempted. Furthermore, it was apparent by then that the Allies had won the war against Germany, and that it probably would only be a matter of months before the war would be over in Europe. Thus, in October 1944 Vannevar Bush had sent a letter to the director of the Radiation Laboratory at MIT asking him to plan the closing of the laboratory, since it was to be dismantled upon cessation of hostilities.

In January 1945 Robert Wilson called a meeting at Los Alamos to discuss "What shall we do about the Gadget?" About forty people attended, including Oppenheimer. One of the topics discussed was whether work should continue on the design of the plutonium bomb, and of atomic bombs in general, since the bombs had been developed mainly for possible use against Germany, and Germany had been essentially defeated. Oppenheimer, by virtue of his position as director of the laboratory and chief scientific adviser to General Leslie Groves, had by that time become involved in matters of military policy and had become familiar with the diplomatic moves that were being contemplated by the United States. He convinced Wilson that possession of atomic bombs was an essential ingredient for the participation of the United States in the projected United Nations that was to replace the League of Nations, and that the new institution had to become an effective instrument for maintaining the peace in a world where atomic bombs were a reality. Oppenheimer had discussed some of these matters with Niels Bohr, who, since coming to Los Alamos in October of 1943, had been deeply concerned about the new conditions created by the development of atomic weapons and had been exploring ways of preventing a nuclear arms race and putting atomic energy under international control.

Similarly, in the spring of 1945, Oppenheimer convinced Edward Teller not to circulate at Los Alamos a petition drafted by Leo Szilard, a member of the Metallurgical Laboratory in Chicago, urging that "the United States would not use the atomic bomb in warfare without warning the enemy." Oppenheimer told Teller,

> in a polite and convincing way, that he thought it improper for a scientist to use his prestige as a platform for political pronouncements. He conveyed to me in glowing terms the deep concern, thoroughness, and wisdom with which these questions were being handled in Washington. Our fate was in the hand of the best, the most conscientious men of our nation. And they had information which we did not possess. Oppenheimer's words lifted a great weight from my heart. I was happy to accept his word and his authority. I did not circulate Szilard's petition.[13]

Ironically, the momentum of the project was such that Oppenheimer would recall later: "I don't think there was any time when we worked harder at the speed-up than in the period after the German surrender [on May 8, 1945] and the actual combat use of the bomb."[14]

By virtue of his directorship of Los Alamos, Oppenheimer became a member of the Scientific Advisory Panel to the Interim Target Committee that was to advise President Truman on the use and control of atomic energy in general, and on the use of the bombs against Japan in particular.[15] The four-member scientific panel, consisting of Arthur Compton, Fermi, Lawrence, and Oppenheimer, recommended using atomic bombs against Japan, finding "no acceptable alternative to direct military use."[16]

In 1947, Fermi was interviewed by FBI agents who were investigating Oppenheimer in connection with his appointment as a member of the U.S. delegation to the United Nations that was to present the Acheson-Lilienthal plan for the international control of atomic energy. Fermi recalled that during the deliberation of the advisory group, "Oppenheimer expressed the wish that it would not be necessary to use the bomb but recognized that its use was a military problem and that the whole purpose of the development of the atomic bomb was its possible use against an enemy nation."[17] Although Bethe was the head of the Theoretical Division at Los Alamos, he was not involved in any of the decisions of where and when to use the bombs.[18]

The Trinity test, to confirm the design of the plutonium bomb with its sensitive implosion mechanism, took place at 5:30 in the morning on Monday, July 16, 1945, in the Jornada del Muerto,[19] roughly halfway between Secorro and Alamogordo in New Mexico. Oppenheimer's reaction at the test was recorded by General T. R. Farrell, Groves's right-hand man:

> Dr. Oppenheimer, on whom had rested a very heavy burden, grew tenser as the last second ticked off. He scarcely breathed. He held on to a post to steady himself. For the last few seconds, he stared directly ahead and then when the announcer shouted "Now!" and there came this tremendous burst of light followed shortly thereafter by the deep growling roar of the explosion, his face relaxed into an expression of tremendous relief.[20]

Immediately after the explosion, many people, including Groves, sought out Oppenheimer to congratulate him. That same morning at Los Alamos, Oppenheimer remarked that his "faith in the human mind has been somewhat restored" and, while greatly relieved that the endeavor had succeeded, he admitted that he was "a little scared of what we had made."[21] He later related that when the bomb detonated he became conscious of the line from the Bhagavad Gita: "I am become Death, the destroyer of worlds." This statement took on melodramatic overtones afterwards. But in fact, the quotation reveals two essential characteristics of Oppenheimer: his erudition and insightfulness, and his inability to hold himself at arm's length from the historic events of which he is part. On the one hand, the quotation captures one of the central principles of Hindu thought, namely, the juxtaposition of oppositions and their interpretations—in this case, life and death; on the other hand, it reveals Oppenheimer, like Krishna, as the embodiment of oppositions.

Bethe had watched the explosion from an observation post some twenty miles from ground zero with Teller, Feynman, and others from the Theoretical Division. He was awestruck and speechless after the bomb went off: "It was too much to say anything."[22]

The incisiveness of the statement that Kenneth Bainbridge had made at the time—"Well, now we're all sons of bitches"—became clear to everyone at Los Alamos after Hiroshima. According to Bethe, the bombing had been eagerly awaited and "people were very happy" that it had happened. But "we were then very much sobered when we got films from Hiroshima, taken the next day, which showed a destruction far beyond anything we had visualized, although we had calculated something very close to it." Philip Morrison and Robert Serber had gone to Hiroshima in early September 1945 and reported back to the Los Alamos scientists the terrible destruction that had been wreaked on the city.[23] Although the earlier pictures had indicated the magnitude of the devastation, Morrison movingly described the enormity of the suffering the bomb had inflicted on the population. After Hiroshima, Bethe and most other scientists from the various parts of the project concluded that a nuclear war must never be waged. This was the message that Morrison had brought back from Hiroshima and Nagasaki. Oppenheimer was perhaps the most troubled by the bombing of Hiroshima and Nagasaki. On August 17 he wrote Henry Stimson, the secretary of war: "The safety of this nation, as opposed to its ability to inflict damage on an enemy power—cannot lie wholly or entirely in its scientific or technical prowess. It can only be based on making future wars impossible."[24]

After the war, both Bethe and Oppenheimer became deeply concerned with the consequences of the atomic bomb. Like many of those who worked on it, they began to lecture extensively to educate the public at large about the dangers and sought to influence Congress to accept placing the bomb under international control.[25]

HYDROGEN BOMBS

Relations between the United States and the Soviet Union deteriorated precipitously after the war. The failure to reach an agreement to place atomic energy under international control, the takeover of Czechoslovakia, the Berlin blockade, the victory of Mao Tse-tung over Chiang Kai-shek on the Chinese mainland, the detection of the detonation of Joe 1, the first Soviet atomic bomb, in late August 1949, and the arrest of Klaus Fuchs in early January 1950 created an atmosphere that led Truman to order the Atomic Energy Commission (AEC) to go full speed ahead with the development of a hydrogen bomb, against the recommendation of the General Advisory Committee (GAC).[26]

After the detonation of Joe 1, the question whether the United States should intensify its effort to develop a Super became the focus of intense debates within the U.S. government. In the fall of 1949, Oppenheimer was opposed to a crash program to build a hydrogen bomb;[27] and in this he was strongly influenced by James Bryant Conant, who in early October had written him that such a bomb would be built "over my dead body."[28]

James Bryant Conant, Vannevar Bush, Frank Jewett, and Karl Compton had organized the National Defense Research Committee

(NDRC) in June 1940, after France fell, in order to mobilize the scientific manpower of the United States for the forthcoming fight. In 1941, when the NDRC became transformed into the Office of Scientific Research and Development (OSRD), Conant became deputy director under Bush, in charge of the S-1 committee that oversaw the making of the atomic bomb. Conant was at first dubious that an atomic bomb could be made to work; but he immediately recognized its devastating potential should it prove achievable, and he was horrified by the prospect. Even after Fermi and his collaborators had built the first successful slow-neutron, self-sustaining pile, Conant still hoped that an atomic bomb wouldn't work. Once it became clear that a U^{235} bomb could in fact be built, he wanted an American monopoly of the weapon, even if this meant the termination of the wartime Anglo-American collaboration on the atomic project. Possession of the weapon would guarantee the imposition of a *pax americana* on the world, since it had become clear to Conant that the European democracies would not be able, after the war, to assume the global responsibility of supporting democratic governments committed to free-market economies.

Consistent with the moral stand he had taken during World War I, Conant believed that atomic bombs should be used during the war if they would ensure an earlier victory. To him there was no difference in killing a hundred thousand Japanese in fire raids using conventional bombs or killing the same number using an atomic bomb. After the war, Conant convinced Henry Stimson, Roosevelt's widely respected Secretary of War who had recommended the use of the bomb to Truman, to write an influential article in the *Atlantic Monthly* justifying the dropping of atomic bombs on Hiroshima and Nagasaki. He supplied him with a draft for the article.

Conant first encountered Oppenheimer early during the war. They met regularly after Oppenheimer had hosted the workshop in Berkeley during the summer of 1942 to study the feasibility of designing a nuclear weapon and after he was named director of Los Alamos. Conant became Oppenheimer's mentor, and a deep bond based on mutual respect developed between the two. It was very probably Conant who convinced Oppenheimer that atomic bombs should be used, without warning, on Japanese cities during the war, and it was Conant who persuaded Oppenheimer in the fall of 1949 to oppose a crash program to develop a hydrogen bomb. Thus, on October 29, 1949, the second day of the meeting of the GAC at which its members formulated their recommendation for the H-bomb program in the aftermath of Joe 1, David Lilienthal, the chairman of the AEC, made the following entry in his diary:

Conant flatly against it [the Super] "on moral grounds." Obviously Oppenheimer inclined that way. Buckley sees no diff[er-ence] in moral question x and y times x, but Conant disagreed—there are grades of morality. . . . Rabi says decision to go ahead will be made; only question is who will be willing to join it. . . . Conant replies: but whether it [the decision] to go ahead will stick depends on how the country views the moral issue.

Conant makes firm point at outset: Can this be declassi-fied—i.e., the fact that there is such a thing being considered, what its effect will be, if it could be made successfully, etc.?[29]

Oppenheimer believed that the atomic bomb had helped prevent further Soviet expansion into Western Europe, but it did not seem necessary to him to have more powerful weapons to deter Russian aggression even if the USSR had atomic bombs. By producing more A-bombs, refining their design to include tactical uses, and having a better delivery capability, he thought the United States would be able to keep its military superiority over the USSR for the indefinite fu-ture.[30] Moreover, when the General Advisory Committee (GAC) of the AEC was confronted with the issue in the fall of 1949, it was uncertain whether a hydrogen bomb could be made. The method then pro-posed had been under theoretical development for seven years, and "in the end turned out to be unpromising, if not useless." Thus, at their meeting of October 28–29, 1949, the GAC concluded: "We all hope that by one means or another, the development of these weap-ons can be avoided. We are all reluctant to see the United States take the initiative in precipitating this development. We are all agreed that it would be wrong at the present moment to commit ourselves to all-out efforts towards its development."

Although it recommended against a crash program, the GAC was not against the exploration of thermonuclear problems. Oppenhei-mer, and five other members of the GAC, made explicit the reason for their recommendation in an annex written by Conant on October 30, 1949, that was attached to the report of the committee. They be-lieved that

the extreme danger to mankind inherent in the proposal wholly outweighs any military advantage that could come from this development. Let it be clearly realized that this is a super weapon; it is a totally different category from an atomic bomb. The reason for developing such super bombs would be to have the capacity to devastate a vast area with a single bomb. Its use would involve the decision to slaughter a vast number of civilians. We are alarmed as to the possible global effects of the radioactivity generated by the explosion of a few super bombs of conceivable magnitude. If super bombs will work at all, there is no inherent limit in the destructive power that may be attained with them. Therefore, a super bomb might become a weapon of genocide. . . .

In determining not to proceed to develop the super bomb, we see a unique opportunity of providing by example some limitations on the totality of war and thus limiting the fear and arousing the hopes of mankind.

Fermi and Rabi, in another annex to the main report, took essentially the same position, expressed it in stronger language, and proposed banning the testing of such a weapon. They asserted that it was "important for the President of the United States to tell the American public, and the world, that we think it wrong on fundamental ethical principles to initiate a program of development of such a weapon." Calling an H-bomb "necessarily an evil thing considered in any light," they recommended a ban on thermonuclear testing: "If such a pledge were accepted even without control machinery it appears highly probable that an advanced state of development leading to a test by another power could be detected by available physical means." They stressed that designing and assembling such a weapon could not be done without tests.

Because Oppenheimer was officially a member of the government by virtue of his chairmanship of the GAC and his membership on numerous Pentagon committees, he could only make broad statements regarding his views. That he was deeply concerned about the threat posed by the Soviet Union and by Communism was well known. He had made explicit his assessment of the Soviet Union in 1947, when addressing some of the issues connected with the control of atomic energy:

We couple with our desire for world order certain minimum conditions about how the United States should continue in its own tradition and continue to work out its own destiny as in the past. And above all, I think, there stands the great conflict with Soviet communism. There may be people who believe that this [system] originated in a desire to provide for the well-being of the people of Russia—for their standard of living, for their health. But whatever its origin, it has given rise to political forms which are deeply abhorrent to us and which we would not only repudiate for ourselves but which we are reluctant to see spread into many areas of the world where there is great lability, as there is in Europe and Asia—in fact, perhaps in most places other than the United States.[31]

But as chairman of the GAC, Oppenheimer could not speak out publicly on the issue of building a hydrogen bomb. Nor was he free to express his strongly felt position against the strategic use of nuclear weapons.[32] Statements about the Super were forbidden under AEC rules after Truman, on January 30, 1950, had made the decision to go forth with a crash program on the Super and had clamped a lid on public discussions by AEC personnel of matters relating to the H-bomb.

On February 12, 1950, in a nationwide television broadcast hosted by Mrs. Roosevelt, on which he appeared with Einstein and Bethe, Oppenheimer could only say that the issues concerned with the building of a fusion weapon

> are complex technical things, but they touch the very basis of our morality. . . .
> It is a grave danger for us that these decisions are taken on the basis of facts held secret . . . wisdom itself cannot flourish and even the truth not be established, without the give and take of debate and criticism. The facts, the relevant facts, are of little use to an enemy, yet they are fundamental to an understanding of the issues of policy.[33]

Some ten years later, Oppenheimer recalled at a seminar his reactions to Truman's decision to proceed with the development of the Superbomb:

> I find myself in profound anguish over the fact that no ethical discussion of any weight or nobility has been addressed to the problem of atomic weapons. . . . What are we to make of a civilization which has always regarded ethics an essential part of human life, and which has always had in it an articulate, deep, fervent conviction . . . a dedication to . . . doing no harm or hurt . . . what are we to think of such a civilization which has not been able to talk about killing almost everybody, except in prudential and game-theoretical terms.[34]

By 1960 Oppenheimer had evidently forgotten that Bethe had forcefully addressed the moral issues at the time. Since Bethe was only a consultant at Los Alamos, after a front-page story in the *Washington Post* on November 18, 1949, disclosed the existence of the program exploring the feasibility of a fusion bomb, he could express his opposition publicly provided no classified information was divulged.[35] In discussions with Victor Weisskopf, George Placzek, and his wife, Rose, he had come to the conclusion that an H-bomb is not really a military weapon: it is primarily a weapon for annihilating large cities. After a war fought with hydrogen bombs, "even if we were to win it, the world would not be . . . the world we want to preserve."[36] And should such a weapon be built, Bethe wanted the United States to pledge that it would never be the first to use an H-bomb. Together with eleven other distinguished physicists, Bethe issued the following statement on February 4, 1950, following President Truman's decision to allow the development of the bomb:

> We believe that no nation has the right to use such a bomb, no matter how righteous its cause. The bomb is no longer a weapon of war but a means of extermination of whole populations. Its use would be a betrayal of all standards of morality and of Christian civilization itself. . . . [W]e urge that the United States, through its elected government, make a solemn declaration that we shall never use this bomb first. The circumstance which might force us to use it would be if we or our allies were attacked by *this* bomb. There can be only one justification for our development of the hydrogen bomb and that is to prevent its use.[37]

Bethe conveyed his passionate opposition to the H-bomb in articles, interviews, and public lectures, especially when Truman, after giving the order for the go-ahead on a crash program on January 31, 1950, had issued an executive order barring all government employees from speaking on the matter. Thus, in the television program conducted by Eleanor Roosevelt, Bethe reiterated his belief that while atomic bombs could still be applied to military targets "hydrogen bombs can only mean a wholesale destruction of civilian populations. We dislike the Russian system because of the means . . . it uses. It has a dictatorship; it suppresses human liberties; it disregards human dignity and human life. We believe in these values. Shall we defend these values by obliterating all Russian cities and their populations?" The only possible reason for the United States to develop a hydrogen bomb was that the Soviet Union might do so before the United States and might then force it to surrender under the threat of its use. And should we develop one, Bethe declared, the United States should unilaterally pledge that "we will never be the first to use hydrogen bombs."[38] In a letter written two days after the broadcast, Bethe informed Victor Weisskopf that Oppenheimer had agreed with what he had said and had "emphasized the necessity of keeping the issue alive and I very much agree with him."

Bethe had initially feared that a public discussion and the airing of the controversy would push the Soviet Union into a race.[39] But his belief that the American public needed to know the facts in order to exercise informed judgment won the day. In influential articles in *Scientific American* and the *Bulletin of the Atomic Scientists* in April 1950, and in numerous lectures, Bethe made his views known.[40]

> I believe the most important question is the moral one: can we, who have insisted on morality and human decency between nations as well as inside our country, introduce this weapon of total annihilation into the world. The usual argument . . . is that we are fighting against a country which denies all the human values we cherish and that any weapon, however terrible, must be used to prevent that country and its creed from dominating the world. It is argued that it would be better for us to lose our lives than our liberty; and this I personally agree with. But I believe that this is not the question; I believe that we would lose far more than our lives in a war fought with hydrogen bombs, that we would in fact lose all our liberties and human values at the same time, and so thoroughly that we would not recover them for an unforeseeably long time.[41]

From the fall 1949 to the spring of 1951—when Stanislaw Ulam and Edward Teller advanced a new mechanism—Bethe's position was that one should only find out whether fusion bombs could be made, how they could be made, and that one should do so without prejudicing later thinking on the production of such weapons.[42] He urged that one assess their cost, in particular what benefit hydrogen bombs have over atomic bombs. In his articles, Bethe repeatedly stressed that very few targets are big enough for a hydrogen bomb, and that one has much more flexibility with a stockpile of smaller atomic bombs. The destruction wrought by a hydrogen bomb is indiscriminate, whereas that caused by atomic bombs could be targeted. Thus, militarily, the H-bomb would not add much to the A-bomb, but A-bombs—particularly tactical ones—can and do add very much to conventional armaments and would constitute a powerful deterrent to a Soviet invasion of Western Europe.

Bethe had additional reasons to oppose a crash program on the H-bomb: too strong an emphasis on its development would prejudice its future disposition. "If this is considered the highest priority development it will automatically follow that it will be built into our military thinking to use this weapon." Moreover, if the USSR and the United States both possessed the hydrogen bomb, the weapon would be a deterrent primarily for the United States because it is more vulnerable than Russia to such an attack, given the locations and clustering of its large cities.

Bethe's involvement in the development of the hydrogen bomb from 1949 until 1954 stands in sharp contrast to his wartime work at Los Alamos. He had been deeply troubled by Stalin's position on free elections in Poland, by the Iron Curtain the Soviet Union had clamped down over Eastern Europe, and by the Soviet Union's direct intervention in Hungary in 1947 and in Czechoslovakia in 1948. He was fully aware that World War III was a real possibility due to Soviet intransigence regarding control of atomic weapons and the future of Germany, its actions in Hungary and Czechoslovakia, and its imposition of the Berlin blockade. Equally important were the American reactions and responses to these Soviet moves—the Truman doctrine, the containment policy outlined in George Kennan's "Mr. X" article in *Foreign Affairs* in 1947, and the sharp increase in the defense budget in 1949. Nonetheless, Bethe was opposed to the development of the hydrogen bomb—*primarily on moral grounds*.[43] To him the most important problem was not how to build a bomb, but whether to build it. In February 1950, he wrote a letter to Morris E. Bradbury, the director of Los Alamos, stating his stand on working to develop a hydrogen bomb:

Dear Dr. Bradbury,

You will probably have heard about my feelings concerning the hydrogen bomb from Carson Marks and from the newspapers. The announcement of the President has not changed my feelings in the matter. I still believe that it is morally wrong and unwise for our national security to develop this weapon. In most respects I agree with the opinions of the General Advisory Committee although I have not seen their report itself. So much has been said about the reasons on both sides that I do not need to go into them here. The main point is that I can not in good conscience work on this weapon.

For this reason, if and when I come to Los Alamos in the future I will completely refrain from any discussions relating to the super-bomb. I have not decided whether this should include work on the [censored]. This will depend essentially on the question of how many problems the super and the [censored] have in common. Therefore on my visits I would primarily concern myself with the problems of the implosion, with problems of neutron diffusion and of efficiency, in other words with classical Los Alamos problems.

It is my feeling that these problems will not be in the center of interest of the Los Alamos Laboratory, particularly of the theoretical division. Therefore it no longer seems necessary for me to plan as many and as prolonged visits as I had been. I am very sorry for that because I enjoy being with you and seeing old friends. However it seems to me the only reasonable conclusion from the present situation, especially considering that there are so many other things I have to do.

But Bethe concluded his letter with the following caveat:

In case of war I would obviously reconsider my position.
With best personal regards to all my friends at Los Alamos,
Sincerely yours,
Hans A. Bethe[44]

With this letter Bethe had essentially dissociated himself from the hydrogen bomb project,[45] and technically was therefore no longer a consultant to it. This allowed him to publish in the April 1950 *Scientific American* his article counseling against the development of the Super.[46] The AEC order barring all employees from discussing any technical aspect of thermonuclear weapons, even if it was not classified, had also specified that "consulting scientists" could not do so.

Ever since Fermi had conceived of a fusion weapon, and throughout the various ups and downs of the investigations by Teller of the feasibility of such a weapon, Bethe steadfastly hoped that the mechanisms that Teller advanced would prove unrealizable. In his testimony to the Joint Committee on Atomic Energy on May 3, 1950, Bethe defended and clarified his position not to work on the development of this weapon.[47] Senator Brian McMahon asked him, given President Truman's decision to initiate a crash program on the H-bomb: "Have you as a citizen the right to interpose your political judgments on the matter and thereby frustrate the contribution that you as a citizen, particularly equipped by Almighty God and the great genius that you have, have you the moral right, I wonder, to withhold?"

Bethe answered: "Well, this a hard question. I think I do. This, after all, is a free country . . . a country which prides itself on giving the right to the individual to decide his own actions." McMahon then challenged him: "But . . . if by possession first of the hydrogen weapon [the USSR] may be able destroy the system that gives you the right of choice, I wonder if you are well advised on insisting on it at this time."

Bethe replied that he had gone through all the political and moral arguments. The fact of the matter was that he firmly believed that the hydrogen "will not do us any good and it will not win the war for us, nor the lack of it will not lose the war for us." At which point Senator Sterling Cole interjected: "This being so, . . . it would seem logical for you . . . to participate in establishing as fact that this weapon is impractical from the military and economic standpoint."

Cole had pointed to the heart of the matter: only by providing his skills and expertise to the government could Bethe gain access to the decision-making process and influence it. Bethe of course knew this, and in his answer he explicitly stated the price he was willing to pay:

> I am not entirely unwilling to participate. As you know, I am a consultant to the Los Alamos Laboratory. I have been there and have advised on these very calculations to establish the feasibility, and amount of material necessary. This is as far as I want to go and I will continue to advise on that. I have advised to the extent of making it more likely to get a successful result. That is, I have said that it would be better to distribute the material in a different manner so that less material will be used. To this extent . . . I want to participate and it is right for me to participate. I do not want to participate in thinking of other ways and means how to construct this weapon.

McMahon then noted that as long as Bethe had that mental attitude he would be "fighting himself all the time," and, great as his genius is, he might not do as good a job because his heart was not in it. Whereupon Cole came to Bethe's defense by pointing out that Bethe's presence at Los Alamos would attract others who might not be so reluctant.

During the summer of 1950 and in January 1951, Bethe did work at Los Alamos on the "classical Super" hoping to prove that it would *not* work.[48] And in fact it did not. But a viable mechanism which would not require large amounts of tritium—which Oppenheimer later called "technically sweet"—was advanced by Ulam and Teller in the spring of 1951, based on their insight "(a) that high densities could be useful and (b) that they could be achieved by a radiation implosion."[49] Bethe then did help to evaluate the proposal, and both Oppenheimer and he supported the development of the weapon at a GAC-AEC conference held in Princeton on June 16–18, 1951. Moreover, Bethe helped design a weapon based on the Ulam-Teller conception. He did so even though he believed at the time that it was "a terrible error" to develop fusion bombs. But since such weapons were now feasible, it was clear that the Russians would also have them, and for that reason the United States had to have them: the balance of terror must not be destabilized, for the avoidance of war depended on it. Even if war were not to break out, the West would be vulnerable to blackmail by the Soviet Union if H-bombs were only in Stalin's hands. In a memorandum entitled "Comments on the History of the H-bomb," written and deposited in the Los Alamos Archives in 1954, Bethe reviewed the history of the H-bomb project. He there stated that "the possibility the Russians might obtain an H-bomb was of course the most compelling argument for proceeding with our thermonuclear program. It was, in my opinion, the *only* valid argument.[50]

In 1968, in an interview with Lee Edson, Bethe recalled the factors that led him to his decision to work on the H-bomb:

> Just a few months before, the Korean war had broken out, and for the first time I saw direct confrontation with the Communists. It was too disturbing. The cold war looked as if it were about to get hot. I knew then I had to reverse my earlier position. If I didn't work on the bomb somebody else would—and I had the thought if I were around Los Alamos I might still be a force for disarmament. So I agreed to join in developing the H-bomb.
>
> It seemed quite logical. But sometimes I wish I were more consistent an idealist.[51]

In February 1952, Bethe went to Los Alamos to help implement the design of an H-bomb. He characterized his role there as follows: "After the H-bomb was made, reporters started to call Teller the father of the H-bomb. For the sake of history, I think it is more precise to say that Ulam is the father, because he provided the seed, and Teller is the mother, because he remained with the child. As for me, I guess I am the midwife"[52]

If there was any resonance in Bethe's statement with Oppenheimer's assertion during his "trial" that, "when you see something that is technically sweet, you go ahead and do it and you argue about what to do about it only after you had your technical success. . . . I cannot very well imagine if we had known in late 1949 what we got to know by early 1951 the tone of our report would have been the same," it was still the case, as Bethe asserted in 1957, that "my inner troubles stayed with me and are still with me and I have not resolved the problem. I still have the feeling that I have done the wrong thing. But I have done it."[53]

Bethe's willingness to work on the bomb did allow him to try to influence policy regarding the testing of the device. It is interesting to note some of Bethe's actions as an "insider" during his stay at Los Alamos in 1952. Since he was now constrained by the AEC guidelines of 1950 regarding public discussions relating to the H-bomb, the only channels of criticism open to him were discussions with the director of Los Alamos and letters to the AEC commissioners. Bethe availed himself of both options when he became concerned about the date set for the first thermonuclear test. He discussed the matter with Norris Bradbury, the director of Los Alamos, and also took pen to hand and wrote Gordon Dean, the chairman of the AEC.[54]

After it had become clear that the Ulam-Teller mechanism would work, plans for a test to take place on Eniwetok, in the Kwajalein atoll, were set in motion. The date of the test was set for November 1, 1952, four days before the presidential elections of that year. From the time he arrived at Los Alamos in February 1952, Bethe became worried that if the test were carried out on that date, if it was successful and this fact became known, it would be used as campaign material. He felt that some politician—from either party—might believe it would help win the election if he referred to such a successful test. Bethe was deeply troubled that such a speech might lead the public to believe "that the accomplishment of a thermonuclear reaction had made us invincible, that we could now take chances in foreign policy and perhaps even risk a major war. It would take a long time to correct this impression, and in trying to do so, we would be unable to use some of the arguments because they are classified."

Bethe was also deeply concerned that a few belligerent, or merely incautious, remarks by some politicians in this country "will play in the hands of Communist propaganda by convincing many otherwise friendly people abroad that this is an important step toward our starting a war. 'Neutralism' will be strengthened, to an extent that it may influence the policy of European governments and become very hard to deal with."[55]

Truman did try to delay the Eniwetok test until after Election Day and keep the issue out of the 1952 campaign. But the AEC accepted the judgment of key advisers that even a brief postponement would be very costly. With Truman's approval, the AEC tried to impose a news blackout on the test. The attempt failed, but the H-bomb never became an issue in the Eisenhower-Stevenson campaign.[56]

The Eniwetok explosion was about eight hundred times more powerful than the Hiroshima bomb. It obliterated Elugelab, an island with an area of over one square mile, and left a huge gaping crater on the ocean floor. In his State of the Union speech in January 1953, in which he informed the nation of the test, Truman indicated that he held out little hope that the arms race in the development and stockpiling of these new superweapons would be contained. Nor did he believe that an agreement on arms control could be negotiated with the Soviet Union in the near future. Rather, he called for a continued tough stand in dealing with "the masters of the Kremlin." His speech did not reveal the sharp disagreement that had existed within the government on whether the Eniwetok test should have been held.[57]

PSAC AND THE NUCLEAR TEST BAN TREATY

A suggestion for a thermonuclear test ban had been put forward by Vannevar Bush in the spring of 1952 before the Eniwetok test had been carried out. Bush was then a member of a high-level State Department Panel of Consultants on Disarmament that had been convened by Secretary Dean Acheson to explore the possibilities of reaching some arms control measures with the Russians. Robert Oppenheimer and Allen Dulles, the deputy director of the CIA, were also on the panel, and McGeorge Bundy served as its executive secretary. The panel came to the conclusion that the United States would be safer if a Soviet-American agreement to ban the testing of an H-bomb could be effected. They urged the cancelation of the forthcoming Eniwetok test in order to allow the next administration—which was expected to be Eisenhower's—to consider a test ban that would bar all future thermonuclear testing. The panel was of the opinion that a delay in the test would not harm the thermonuclear program. But holding the test would destroy all prospects for an agreement, since the United States would have obtained important information from the explosion. Oppenheimer strongly supported the panel's recommendations and indefatigably worked for their adoption.[58] The proposal never reached Truman's desk, since Acheson had killed the plan.[59]

Though it had seemed utopian to some, Bethe had thought that the October 1949 Fermi-Rabi proposal to explore the possibility of a thermonuclear test ban was realistic. In 1952, while working on the H-bomb project, Bethe was unaware of the Bush plan. When later, as a member of PSAC, he became acquainted with it, he strongly supported the premise of the proposal for it might have prevented—*at that time*—the escalation of the arms race with fusion weapons.[60] An Ulam-Teller hydrogen bomb was an extremely complicated device that could not be built without testing it. Bethe had hoped—even while working on the bomb—that an agreement might be reached with the Soviet Union that would allow theoretical exploration and understanding, but would forbid steps toward any major "hardware" development and the testing of a weapon. As had been emphasized by Fermi and Rabi, the violation of such a test ban by the USSR could easily be detected, and the United States could then test its own device shortly after the recognition of such a violation—probably within a year. Bethe felt that this was quite an acceptable risk because the superior arsenal of A-bombs would actually have protected the United States adequately. In a letter to Gordon Dean in late May 1952, Bethe warned him that "if we now publicly intensify our efforts we shall force the Russians even more into developing the weapon which we have every reason to dread."[61] And as he commented two years later, "It is possible, perhaps likely, that the Russians would have refused to enter an agreement on this matter. If they had done so, this refusal would have been a great propaganda asset for us in the international field and would in addition have gone far to persuade the scientists of this country to cooperate in the H-bomb program with enthusiasm."[62] The publication of Andrei Sakharov's *Memoirs* made it clear that indeed the Soviets would not have agreed to such a proposal.[63]

In 1949 and in 1951 Bethe felt strongly that an H-bomb should not be developed; and in 1954 he called its development a "calamity."[64] He still is of the opinion that it should not have been built. Thus in 1983 he declared that "it was unnecessary. It should not have been done. We would now be very much better off if the hydrogen bomb had never been invented, because the thousands of missiles could carry only low-yield weapons. They now carry high-yield weapons. The world would be a far less dangerous place if the H-bomb had never been invented."[65]

During the 1980s the planet was indeed a dangerous place. The two superpowers were confronting one another with tens of thousands of nuclear weapons, and a rhetoric of "nuclear tipped weapons" was being deployed to minimize their dread and to suggest that their use was not inconceivable nor recovery after their use impossible. Although there has been some progress in reducing the number of warheads and missiles since the end of the Cold War, Bethe's statement still holds true. In remarks made in 1988 during a visit to the United States following his rehabilitation by Gorbachev, Sakharov echoed these sentiments.

> While both sides felt that this kind of work was vital to maintain a balance, I think that what we were doing at the time was a great tragedy. It was a tragedy that reflected the tragic state of the world that made it necessary, in order to maintain peace, to do such terrible things. We will never know whether it was really true that our work contributed at some period towards maintaining peace in the world, but at least at the time we were doing it, we were convinced this was the case.

The world has now entered a new era, and I am convinced that a new approach has now become necessary. And I think in each case when a person makes a decision, he should base that decision on an absolute conviction of his rightness, and only under such circumstances can we ever find mutual understanding. And under such circumstances and in doing so, it is very important, and essential in fact, to find out all points of difference as well as the points of coincidence and the points where the views are the same.

Bethe's actions regarding work on the H-bomb illuminate deeply held convictions. Although Bethe has always maintained that scientists should feel responsible for the consequences of their work, and that they should consider these before starting to work on potentially hugely destructive weapons such as atomic and hydrogen bombs, he did not believe that—within the context of the Cold War—they *collectively* had the right to decide to refuse to work on such weapons, for "it would set the scientific community up as a superpolitical body. . . . Setting scientists up as the sole judges of their actions means, in the extreme, condoning treason such as that of Klaus Fuchs." A scientist of course has the right, and perhaps the obligation, to judge the conduct of the state which reaps the benefit of his work. And should he find himself opposing the internal or external politics of his country, he has of course the right to indicate his convictions and the right to act on them as an individual by withdrawing his services. Bethe feels strongly that it is the duty of scientists—individually and collectively—to make their opinion and vision known to the government; but that it is for the government, and not the scientist, to make the final decision.

Bethe believes that scientists have a vital role to play in the decision-making process of the state, but that in order to fulfill this function "scientists (at least some of them) must be willing to work for the government and in the government, and they must be willing to work on weapons." Moreover, it seemed to Bethe "that by working on these weapons one earns the right to be heard in suggesting what to do about them."[66]

This is the modern form of Kant's contract with Frederick II. How can the use of reason take the public form that it requires, how can *sapere aude* be exercised publicly, while individuals are obeying privately as scrupulously as possible? The solution that Bethe proposed—and that he acted upon—is a presumed contract with the state, a contract of rational authority with free reason: the public and free use of autonomous reason will be the best guarantee of loyalty and allegiance, on condition, however, that the political principle that must be obeyed is itself in conformity with reason.

Throughout his life, Bethe has acted on these convictions. Thus, although he did not believe that a weapon such as the H-bomb ought to be built and he disagreed with the decision taken by the government, Bethe did not sever his ties with Los Alamos in late 1949 after Truman had ordered a crash program. Furthermore, given the context of the Cold War and his assessment of the character of Stalin's dictatorship and of the nature of his totalitarian state, he went to work at Los Alamos to help in the development of a bomb based on the Ulam-Teller mechanism after its feasibility had been demonstrated.

It is indicative of Bethe's constant grappling with moral issues that he now advocates, and has been urging, that fellow scientists *collectively* take a Hippocratic oath not to work on designing new nuclear weapons, given the changed context brought about by the collapse of the Soviet Union. Thus, on the occasion of the fiftieth anniversary of Hiroshima in 1995, he went to Los Alamos to address the scientists there and convince them that one should not work on the further improvement of nuclear weapons. He then issued the following statement:

> As director of the Theoretical Division of Los Alamos, I participated at the most senior level in the World War II Manhattan Project that produced the first atomic weapons.
>
> Now, at the age of 88, I am one of the few remaining such senior persons alive. Looking back at the half-century since that time, I feel the most intense relief that these weapons have not been used since World War II, mixed with the horror that tens of thousands of such weapons have been built since that time—one hundred times more than any of us at Los Alamos could ever have imagined.
>
> Today we are rightly in an era of disarmament and dismantlement of nuclear weapons. But in some countries nuclear development still continues. Whether and when the various Nations of the world can agree to stop this is uncertain. But individual scientists can still influence this process by withholding their skills.
>
> Accordingly, I call on all scientists in all countries to cease and desist from work creating, developing, improving and manufacturing further nuclear weapons—and, for that matter, other weapons of potential mass destruction such as chemical and biological weapons.
>
> Hans Bethe[67]

He went to Los Alamos to tell people there that they should only work on the maintenance of a much reduced stockpile and to desist "from applying nuclear weapons to new purposes." The anniversary of the bombing of Hiroshima offered him the possibility of stating his views publicly—fully aware of the ineffectiveness of "individual action."[68]

As is well known, Oppenheimer's opposition to a crash program for the development of the hydrogen bomb, and his support of explorations to ascertain whether the USSR would be willing to conclude an agreement that neither side develop the H-bomb, eventually led to the revocation of his clearance and his "trial."[69] In a famous 1954 *Harper's* article whose title was—like Emile Zola's earlier one in defense of Captain Dreyfus—"J'accuse," Joseph and Stewart Alsop identified the various kinds of "Oppenheimer haters" responsible for orchestrating Oppenheimer's fall. High level government officials were involved, men such as AEC chairman Lewis Strauss, whose exaggerated concern with security Oppenheimer had ridiculed before the Joint Atomic Energy Committee in June 1949, and William Borden, the Joint Atomic Energy Committee's executive director who believed Oppenheimer to be disloyal and responsible for delaying the building of a hydrogen bomb. There were also the "air force zealots"—air force generals like Curtis LeMay and Roscoe Charles Wilson who were "dedicated" to strategic bombing with "big bombs," H-bombs in particular. In addition, there were scientists like Edward Teller and Wendell M. Lattimer, the associate director of the Radiation Laboratory at Berkeley, who opposed Oppenheimer because of his character and the influence he wielded.

Sociologist Philip Rieff has remarked that the Oppenheimer hearings gave proof of something more sinister than the Alger Hiss trial, because "if Hiss was condemned for acts and admired for his strong character (set off against the neurotic Chambers who had accused him), Oppenheimer was condemned for his character and praised for his actions."[70] The Gray Board found Oppenheimer's "susceptibility to influence" a threat to security. Although the Board did not doubt Oppenheimer's loyalty, it denied his appeal to have his security clearance reinstated because of his lack of "enthusiastic support" of the security program and for a crash program to develop a hydrogen bomb. The Board also expressed concern with "his highly persuasive influence in matters in which his convictions were not necessarily a reflection of technical judgment" nor "necessarily related [to the] strongest offensive military interests of this country." In arriving at this position the Board had accepted the contention that the only viable strategy was one based on massive retaliation with hydrogen bombs, in contrast to Oppenheimer's commitment to a more balanced, defensive strategy that relied on tactical atomic weapons and an air defense system.

With the loss of his security clearance, Oppenheimer's life changed. On the one hand, a great responsibility had been lifted and he could now cultivate his wide interests and devote more time and energy to his duties as director of the Institute for Advanced Study. He remained an avidly sought after speaker and devoted these public occasions to address a deeply troubling problem: With what shall we replace the program of the Enlightenment? In what ways shall we reconstitute our lives when our faith in rationality and in progress have been undermined? Neither rationality nor religion had prevented the genocides of World War II, and the irrationally huge arsenals of fission and fusion bombs that could obliterate life on the planet made for a precarious peace at best. The vision and promise of collective well-being of Marx and Engels had been transformed into a totalitarian dictatorship whose excesses rivaled those of Hitler's national socialism, and at home the democratic foundations had been threatened by the demagoguery of Senator Joseph McCarthy, by an anti-Communist crusade, and by the formation of the national security state. Oppenheimer's answer to these questions will be considered in the next chapter.

Bethe, too, addressed these same questions. He was deeply troubled by the irrationality and excesses of the nuclear arms race and by the threats posed to moral standards and to democratic institutions by the loyalty oaths; by the witch hunts of colleagues like Philip Morrison; and especially by the "trial" of Robert Oppenheimer, which distressed him profoundly. His response to the threats these challenges posed was in part shaped by his modest attitude. He tried to be effective under the given conditions and their constraints by trying to effect changes that would bring about a more rational world. While after 1954 Oppenheimer was excluded from the national corridors of power, Bethe became more and more involved in policy issues relating to national security.

The successful Soviet launch of Sputnik I, the first man-made satellite, on October 4, 1957, caught the United States by surprise. It raised the fear that the USSR had overtaken the United States in its military technology. The immediate reaction of President Eisenhower was to summon to the White House James Killian, the president of MIT together with a group of scientists that included Bethe and I. I. Rabi. His goal was to elicit their assessment of the space and defense programs and to obtain their advice on how to strengthen American science and technology. At that meeting, Rabi argued forcefully that the president needed an "outstanding full-time scientific adviser" on his staff. Killian concurred and suggested the creation of a small commit-tee of scientists to advise him on crucial scientific and technological issues, on the model of the Council of Economic Advisers. On November 7, in a nationwide address, Eisenhower announced the appointment of Killian to the new post of special assistant to the president for science and technology, and that the existing Science Committee of the Office of Defense Mobilization (ODM) would be enlarged and moved to the White House. By December 1, the ODM Science Advisory Committee had been reconstituted with Killian as chairman and renamed the President's Science Advisory Committee (PSAC). The new members included Robert Bacher, William O. Baker, Bethe, James H. Doolittle, James B. Fisk, George B. Kistiakowsky, Edwin Land, Edward M. Purcell, H. P. Robertson, Jerome Wiesner, and Herbert York. As Killian was to remark: "The committee served under highly advantageous conditions for a top level advisory group, conditions that were almost unique to the Eisenhower [and Kennedy] years."[71] The tone of PSAC was set by people like Kistiakowsky, Purcell, and Bethe—men whose character and experience had tempered any excessive sense of elitism or intellectual arrogance. "They knew how to mix without condescension or shyness with high political, military and industrial personnel. . . . They loved science and wanted others to share their enthusiasm for it and to discover its power to make men and women a little more creative, a little more objective, and a little more humane."[72] The recommendations of PSAC were collective recommendations that had the support of all its members. PSAC was given free access to the president through Killian, and thus gave scientists who opposed the stand taken by Edward Teller, Ernest Lawrence, and Lewis Strauss on nuclear testing and disarmament direct contact with Eisenhower. In addition, PSAC secured the presence of new viewpoints in the internal debates over the administration's policies in these matters.

To carry out PSAC's duties, Killian formed several panels that were to address specific problems. Each panel was made up of regular PSAC members augmented by scientists and engineers invited from outside and was usually chaired by a PSAC member.[73] One of the first panels to be constituted addressed arms control and the possibility of a moratorium on the testing of nuclear bombs.

Starting in 1950, but especially after the massive thermonuclear tests that the United States[74] and the Soviet Union were carrying out, worldwide public reaction against atmospheric nuclear-weapons testing with its extensive radioactive fallout became outspoken, impassioned, and vehement. Shortly after the launching of Sputnik, the Soviet Union sought to capitalize on this success by again making proposals—some of them self-serving—regarding disarmament. One of these was for a moratorium on nuclear testing for a period of two or three years to start on January 1, 1958. The Soviets gained a great propaganda victory with this proposal.

At the first meeting of PSAC in early December 1957, Caryl P. Haskins, the president of the Carnegie Institution in Washington, and Rabi proposed that a panel on disarmament be established. The suggestion was accepted, and a panel on disarmament was organized under the chairmanship of Haskins with Bethe, Rabi, Herbert Scoville of the Central Intelligence Agency, and Herbert York as members. At the first meeting of the disarmament panel on December 29, 1957, Bethe recommended that it address the issue of a test ban, in particular that it make a technical study of the impact of a test ban on United States and Soviet weapons programs and of the feasibility of monitoring a suspension of nuclear weapons tests. Bethe was primarily interested in a test ban as a step toward disarmament, with radioactivity in the atmosphere only a secondary concern. The panel endorsed Bethe's idea. Killian informed the National Security Council at its meeting on January 6, 1958, of Bethe's suggestions, whereupon Eisenhower asked the National Security Council (NSC) to sponsor the technical study on detecting nuclear tests. Killian then recommended that an interagency committee, with Bethe as chairman, be set up to investigate the feasibility of detecting violations if a moratorium on nuclear testing were to be declared by the United States.[75]

Eisenhower was eager to implement the idea of a test ban. He fervently believed that it was wrong for the United States to view the issue of a test ban negatively, for he could not conceive a long-term solution to the danger of nuclear weapons without first establishing a test ban. Already in April 1954, a month after the "Bravo" test with its monstrous fallout and calamitous effect on the crew of the *Lucky Dragon*, a Japanese fishing vessel, he had asked Lewis Strauss to study the matter. John Foster Dulles and the committee that looked into the matter came back with a report that was nearly unanimous in opposing a test moratorium. Eisenhower nonetheless continued exploring the possibility of curtailing the buildup of nuclear weapons and the feasibility of a test ban, and appointed Harold Stassen as his disarmament adviser. The matter of a nuclear test ban treaty with Soviet Russia became an issue in the 1956 presidential election when Adlai Stevenson announced his favoring one. The issue again assumed national prominence in the fall of 1957 with the opening of the twelfth General Assembly of the United Nations at which the Soviets formally called for an agreement on a nuclear test moratorium.

In a January 12, 1958, letter to Nikolai Bulganin, the Soviet prime minister, Eisenhower suggested that the Soviet Union join the United States in technical studies of the possibility of verification and supervision of disarmament and test ban agreements—an essential requirement before the United States would sign any agreement. In fact, following the January 6 meeting of the NSC, Killian and Robert Cutler, who handled national security affairs for Eisenhower, had selected the members of the interagency technical disarmament studies committee that Bethe was to chair. The committee became known as the Bethe Panel. Eisenhower was fully aware that certain very influential persons—in particular Lewis Strauss, Edward Teller, and Ernest Lawrence—vehemently opposed any test ban and were insisting that only an infallible on-site detection inspection system could guarantee that the Soviets were not cheating. The Bethe Panel was to provide the technical assessment to confront the arguments against any proposals for a test ban agreement that might be considered. The panel undertook to give answers to the following questions: Could both atmospheric and underground Soviet nuclear tests be detected? What was the composition of the Soviet and the U.S. nuclear arsenals, and how did their strengths compare? What constraints would a nuclear test ban impose on the Atomic Energy Commission's weapons laboratories?[76] Killian summarized Bethe's subsequent contribution as follows:

> Bethe . . . was to become one of the heroes of the long campaign that led to the test ban of 1963. . . . Bethe possesses a grave nobility of character that has commended the respect and affection of all who have worked with him.
> With these qualities and his deep knowledge of nuclear matters, Bethe gained the confidence of the interagency committee and directed its work with skill . . . the members worked well together, and some even changed their views during the study.[77]

But given the nature of its composition, the Bethe Panel reached only modest conclusions. In a report issued in late March 1958, the panel described "a practical detection system" that would identify nuclear explosions in the Soviet Union except for very low-yield underground detonations. This detection system would require observation stations, mobile ground units, and the right to fly over Soviet territories. The report conceded that a ban would lead to some deterioration of the weapons laboratories and that the United States might benefit from additional testings of "clean," small, and relatively inexpensive nuclear weapons. The report also indicated that the panel could not decide whether a test ban would be to the net military advantage of the United States. This assessment in turn left PSAC free to make its own judgment of the comparative consequences of a test ban.

The report of the Bethe Panel led PSAC to recommend that the United States sign a nuclear test ban treaty after a reliable test detection system had been designed. A conference by technical experts from the United States, Great Britain, and the Soviet Union was held in Geneva during the summer of 1958.[78] Bethe was arguably the most influential and respected participant at the conference. He spoke rather frequently even though he was only an adviser to the American delegation. He was not an official delegate because Lewis Strauss had objected; Strauss had wanted Teller to be one of the delegates, but he had been disqualified because of his vigorous support of testing.[79] The conferees issued a report indicating that a control system acceptable to all parties was possible. It recommended that 180 seismic stations be established and that tests with a yield greater than 5 kilotons would call for on-site inspection.[80]

The work of the Bethe Panel and of the PSAC disarmament committee culminated in Moscow in July 1963, with the initialing by representatives from the United States, Great Britain, and the Soviet Union of a treaty banning nuclear weapons tests in the atmosphere, in outer space, and under water. The treaty had a most welcome environmental effect as it eliminated most atmospheric tests and their radioactive fallout. However, it exerted but little effect on the nuclear arms race since it allowed unlimited underground testing. Yes it was a crucial first step.

The road to the signing of this (partial) nuclear test ban treaty in the summer of 1963 was rocky. That a treaty was signed is one of Bethe's finest achievements. He considers his efforts in ending the era of extensive atmospheric testing, and the role he played in having issues of arms reduction discussed and taken seriously at the highest levels of the U.S. government, his most valuable and important contributions as a scientist in the political arena. And his assessment takes into account the influential role he has played in the debates over anti-ballistic missiles, the Strategic Defense Initiative, energy, and nuclear energy.

6.

ON SCIENCE AND SOCIETY

> We live in an unusual world, marked by very great and
> irreversible changes that occur within the span of a
> man's life. We live in a time where our knowledge and
> understanding of the world of nature grows wider and
> deeper at an unparalleled rate; and where the problems
> of applying this knowledge to man's needs and hopes
> are new, and only a little illuminated by our past history.
> —J. Robert Oppenheimer (1962; cited
> in Oppenheimer 1984, 123)

On November 25, 1947, Oppenheimer delivered the second Arthur D. Little Lecture to a packed audience of over a thousand that had gathered in Walker Memorial Hall at MIT.[1] The purpose of the lectureship was to promote "interest in and stimulate discussion of the social implications inherent in the development of science" by "secur[ing] the record of the deepest thoughts and convictions of [the] lecturers, based on [their] many years of experience in their contact with science, society, government, economics, and the humanities."[2] Oppenheimer was surely qualified to accept the invitation to speak on the subject of the lectureship. He had sprung into national prominence for his "genius and leadership"[3] as director of the Los Alamos Laboratory and had been a member—and the preeminent scientific adviser—of the Lilienthal board that had drafted the American proposals for the International Control of Atomic Energy in 1947. He was then also chairman of the General Advisory Committee (GAC) to the Atomic Energy Commission and had just assumed the directorship of the Institute for Advanced Study in Princeton.

Oppenheimer had entitled his lecture "Physics in the Contemporary World," but at the outset he commented that he did not mean to imply by it "an over-estimate of physics among the sciences, nor too great a myopia for these contemporary days." He was going to speak *about* physics rather than *of* physics and more, specifically, address the general and difficult topic of the relation between science and civilization. That he could do so, he explained, reflected, on the one hand, his self-consciousness as a physicist who had "been forced to become aware of what it is that [he is] doing," and, on the other, "the experiences of this century, which have shown in a poignant way how much the applications of science determine our welfare and that of

178

our fellows, and which have cast in doubt that traditional optimism, that confidence in progress which have characterized Western Culture since the Renaissance."[4]

Oppenheimer indicated that since the cessation of the hostilities, physics in the United States was once again thriving, in fact "booming," and illustrated some of the progress that was being made. But he noted that the physicists labored while bearing a cross, for they "felt a peculiarly intimate responsibility for suggesting, for supporting, and in the end, in large measure, for achieving the realization of atomic weapons. . . . In some sort of crude sense which no vulgarity, no humor, no over-statement can quite extinguish, the physicists have known sin; and this is a knowledge they cannot lose." However, even though from science may come, as in this last war, "a host of instruments of destruction which will facilitate that labor, even as they have facilitated all others,"[5] this result is no reason to refrain from doing science: "No scientist, no matter how aware he may be of these fruits of his science, cultivates his work, or refrains from it, because of arguments such as these. No scientist can hope to evaluate what his studies, his researches, his experiments may in the end produce for his fellow men, except in one respect—if they are sound, they will produce knowledge." Oppenheimer then queried whether "there are elements in the way of life of the scientist which need not be restricted to the professional, and which have hope in them for bringing dignity and courage and serenity to other men. Science is not all of the life of reason; it is part of it. As such what can it mean to man?"

The answer he gave was affirmative, and he submitted that the elements are to be found in the way scientists have organized their community. It is a community that supports skepticism,[6] in which there is "great caution in all assertions of totality, of finality, or absoluteness." The fact that scientific communities had fostered a total lack of dogmatism and authoritarianism implied that the practitioners had become "enlightened" in the Kantian sense and had taken leave of their self-caused immaturity. Could one extrapolate from the small, the scientific community, to the large, the civic polity? Oppenheimer did not answer the question but asserted that "science alone has turned out to have the kind of universality among men that the times require."[7] He apologized for not answering the question he had posed but hoped that, like Bertrand Russell on an earlier occasion, he had left "the vast darkness of the subject unobscured."

As time went on, he tried to formulate an answer. The relation of science to culture became a recurring theme of Oppenheimer's public addresses after World War II. Thus, in a lecture delivered at Vassar College in October 1958, he posed the question whether the remarkable growth of science could go on and whether it should. The answer that he gave to the first question was that it seemed clear to him that this growth would continue, "unless very bad things indeed happen to stop it." Whether it should go on "comes directly to the interplay . . . between knowledge, between science and the good." He then stated:

> There is an easy way of saying that the relation is obvious and almost one of identity. There is, for instance, one simple thing: people who practice science, who try to learn, believe that knowledge is good. They have a sense of guilt when they do not try to acquire it. This keeps them busy. They think that the correction of error is a virtue; they live in a kind of dedication to this ideal. I do not mean live it twenty-four hours a day, but insofar as they are working in science, this is what it is all about. I happen to share this. It seems hard to live any other way than thinking that it was better to know something than not to know it; and the more you know, the better, provided you know it right, provided you know it honestly in its depth and fullness. And this is surely the only clear simple answer to the question that is so often asked, that is in itself almost ridiculous, "What is the responsibility of the scientist?" It is to remain thus dedicated.[8]

In the Vassar lecture, he returned to the question he had left unanswered in his MIT lecture of 1947:

> Our life in science rests on a commitment to the value of learning. It is often said that this is all that is needed as a basis for a good life. Because science rests on this value, therefore human life can be based on it; therefore social institutions can be based on it: and therefore the problem of finding an ethic for today is solved. I am very little persuaded. It seems to me much too narrow, much too narrow even for men of science . . . to do justice to the breadth of human feeling; it does not even account for what most scientists do much of their time. But it does show one thing, that even this activity of increasing human knowledge, which is concerned with truth, is a human activity which without the human dedication to something which is not in any abstract sense necessary, which in no way is itself implied by knowledge, which could in some sense perfectly well disappear from human society without a belief in its worth, would as a human activity dry up and perish.[9]

The "ramification" of science brought about by its amazing growth and the ensuing specialization is a subject Oppenheimer addressed repeatedly in the late 1950s. Thus, in 1959, in a contribution to the Congress for Cultural Freedom, Oppenheimer asserted that although analogies—especially formal, mathematical analogies—can be found between branches of science,

> there is no logical priority of any one science over any other. The behavior of living matter cannot be deduced from the laws of physics. There is simply an absence of contradiction. The criteria of order, harmony and consistency, which are as important in every discipline as accurate observation and correct reasoning, differ from one science to another. . . . To all this must be added the sense of imperfection, of contingency, of endlessness, which arises from the study of nature.[10]

Other recurring themes in his later speeches are the relation between individual and community, between tradition and discovery, and the "relation between the explosive growth and the great success of the sciences on the one hand, and the quality, nature, of our public discourse, our common discourse, and the public sector of our lives on the other." He decried the gulf that had emerged between the discourse of the sciences and that of the culture at large,[11] for

> "it is communication that makes men of us. It is communication that makes possible culture, to live a civilized life, to teach our children, and to learn from our parents and to help one another.
> Communication between the generations, communication between different walks of life have taken on a difficulty that is quite unprecedented; and this calls for patience, for teaching, and for vigor in a way that has never faced man before.[12]

The chasm between the sciences and the culture at large could only be bridged through education. What Oppenheimer said in 1956 about the teaching of physics on the occasion of the twenty-fifth anniversary of the founding of the American Institute of Physics applies to all the sciences:

> Physicists not only invent and discover; they also explain. . . .
> Every scientific advance, past or contemporary, has two traits; it is an enrichment of techniques; it enables us to do what we could not do before, or to do it better: it is know-how. It is also, on the other hand, the answer and reformulation of questions long agitating man's curiosity, something to contemplate, in Peirce's words, "the demi-cadence which closes a musical phrase in the symphony of our intellectual life," a glimpse of harmony and order, a thing of beauty: it is knowledge. We tend

to teach each other, except in the golden years of graduate and postdoctoral study and apprenticeship, more and more in terms of mastery of technique, losing the sense of beauty and with it the sense of history and of man. On the other hand, we tend to teach those not destined to be physicists too much in terms of the story and too little in terms of substance. We must make more humane what we tell the young physicist, and must seek ways to make more robust and more detailed what we tell the man of art or letters or affairs, if we are to contribute to the integrity of our common cultural life.[13]

But his talk about community and communication had also another facet. In a lecture entitled "Prospects in the Arts and Sciences," delivered shortly after his trial, Oppenheimer gave his most anguished statement on the human condition and his own "aching sense of great loneliness." The artist and the scientist depend

on a common sensibility and culture, on a common meaning of symbols, on a community of experience and common ways of describing and interpreting it. He need not write for everyone [or paint or play for everyone]. But his audience must be man, and not a specialized set of experts among his fellows. Today this is difficult . . . the community to which he addresses himself is largely not there; the traditions and the culture, the symbols and the history, . . . the common experience, . . . have been dissolved in a changing world.

EPILOGUE

What someone is, begins to be revealed when his talent
abates, when he stops showing us what he can do.
—Nietzsche

There is a certain irony in tracing the parallel lives of Bethe and Oppenheimer. During the 1930s, when Oppenheimer was creating his great school of theoretical physics in Berkeley, he was the center of an active social and intellectual community. When he became active in political circles in the aftermath of the Spanish Civil War, the significance of community was enlarged for him. Los Alamos further expanded the meaning of community. But the change in the interior landscape that T. S. Eliot so movingly evoked in the *Journey of the Magi* also describes insightfully Oppenheimer's experience at Los Alamos and Alamogordo.

> This set down
> This: were we led all that way for
> Birth or Death? There was a Birth, certainly,
> We had evidence and no doubt. I had seen birth and death,
> But had thought they were different; this Birth was
> Hard and bitter agony for us, like Death, our death.
> We returned to our places, these Kingdoms,
> But no longer at ease here, in the old dispensation,
> . . . .[1]

Stafford Beer, that wise analyst of the human scene who was deeply affected by the cybernetics revolution, observed that World War II demonstrated that *Homo faber*, man the maker of things, needed to be replaced by a new kind of man, *Homo gubernator*, the steersman of large complex, heterogeneous, interactive systems.[2] Oppenheimer discovered himself as that new kind of man at Los Alamos. He quickly acquired an insight into and an understanding of how complex interactive systems function, what makes them viable, and what kind of control and organization they require. After the war, the corridors of power in Washington offered him the scope to nurture these capabilities.

Bethe, on the other hand remained the supreme version of *Homo faber* and grew into the role of *a moral Homo faber, a Homo faber with a conscience*. During the 1930s, he tended to be self-sufficient and somewhat of a loner—both socially and intellectually. He, too, became transformed by Los Alamos. After the war he became the person most responsible for creating the remarkable community formed by the members of the Newman Laboratory and the physics department at Cornell and set its moral and scientific standards. He has become its beloved patriarch. In contrast, Oppenheimer became a lonely and somewhat solitary personage. Whereas Bethe found comfort in the community he helped cement at Cornell—a community molded by his Los Alamos experience—Oppenheimer, perhaps because of the unique role he played at Los Alamos and the special responsibility he bore for its success, could not replicate its spirit in Princeton. The solutions he sought became individualistic. Furthermore, giving up scientific research after World War II robbed him of the self-confidence necessary to act as a moral leader.

Bethe, on the other hand, became a moral force. Bethe as a moral actor offers an interesting contrast to Bethe the producer of scientific knowledge. Bethe's greatness as a scientist is the result of his exceptional powers of synthesis, his remarkable facility to collect, master, and use the intellectual tools and products of the scientific communities to which he belongs, and his singular ability to delineate the problems he undertakes to solve in simple terms that still capture the essentials of the issue. If we exclude his review articles, most of Bethe's scientific productions would probably also have been carried out by others, in more or less the same way, and, except for a few notable instances, at more or less the same time. But it is very likely that it would have taken them much longer than Bethe to do so, and it probably would have taken the efforts of more people.

I am suggesting that in his own scientific productions, Bethe has helped to stabilize the conceptual tools of the community. It is within the moral and the political spheres that Bethe has been able to overcome the constraints of community, to break the resistances offered by conventionality, intimidation, inertia, and lack of courage and to act courageously in the interests of community and humankind. When Bethe assesses that the work of someone else provides useful new tools and valuable new insights into problems, he will go out of his way to have the community recognize the value of the approach no matter what the resistance—and this both in the scientific and in the political sphere. For example, Bethe was Feynman's most forceful supporter at the time when Feynman's approach to quantum field theory was ill-understood and seemed to have no rigorous justification. The story was repeated with Freeman Dyson, and later with Kenneth Wilson, and also on numerous less portentous occasions.

Within the political sphere, Bethe's archives are replete with instances of his taking forthright stands defending the positions that he considered right and attacking those he believed would undermine our democratic institutions or would tear the fragile fabric that holds universities together—this at precisely those times when it required courage to do so. Thus he chastised Oppenheimer for his behavior concerning Peters; he zealously defended his colleague Philip Morrison against the accusations of some members of Cornell's Board of Trustees; he was among Oppenheimer's staunchest defenders when his loyalty was impugned; and in 1969 he was the author of a faculty report defending academic freedom when Cornell was beleaguered by radical students and faculty members. There are many more examples involving his public stand on issues such as the H-bomb, test ban treaties, disarmament, and the Strategic Defense Initiative (SDI).

Making it possible for Bethe to act so decisively as a moral agent are his self-confidence, the force of his personality, and the support of the community he has helped to create at Cornell. All these were secured and are nurtured by his powers and his continuing creativity and accomplishments as a scientist. His scientific labors are his anchor in integrity. Neither that anchor nor such a community was available to Oppenheimer after Los Alamos.

Yet, nowhere are the affinities that bound these two men voiced more clearly than in the lecture by Oppenheimer that was broadcast on Christmas Day 1954 entitled "Prospects in the Arts and Sciences."[3] It had been recorded shortly after the Gray Board and the Atomic Energy Commission had upheld the revocation of his clearance. The new world that Oppenheimer had described in earlier public lectures seemed bleaker than ever. But the lecture was also a guideline on how to comport oneself in this new world:

> This is a world in which each of us, knowing his limitations, knowing the evils of superficiality, and the terrors of fatigue, will have to cling to what is close to him, to what he knows, to what he can do, to his friends and his tradition and his love, lest he be dissolved in a universal confusion and know nothing and love nothing. . . . [And] if we must live with a perpetual sense that the world and the men in it are greater than we and too much for us, let it be the measure of our virtue that we know this and seek no comfort. Above all, let us not proclaim that the limits of our powers correspond to some special wisdom in our choice of life, or learning, or of beauty.[4]

The new world was infinite and

this perpetual, precarious, impossible balance between the infinitely open and the intimate, this time—our twentieth century—has been long in coming; but it has come. It is, I think, for us and our children, our only way.

This is for all men. For the artist and for the scientist, there is a special problem and a special hope. . . . Both the man of science and the man of art live always at the edge of mystery, surrounded by it; both always, as the measure of their creation, have had to do with the harmonization of what is new with what is familiar, with the balance between novelty and synthesis, with the struggle to make partial order in the total chaos. They can, in their work and in their lives, help themselves, help one another, and help all men. They can make the paths that connect the villages of arts and sciences with each other and with the world at large the multiple, varied, precious bonds of a true and world-wide community.

This cannot be an easy life . . . but this, as I see it, is the condition of man; and in this condition we can help, because we can love, one another.[5]

In this lecture, Oppenheimer had returned to Felix Adler's call to strengthen the bonds between *all people* and to deepen their interconnectedness. *Bethe has lived the life that Oppenheimer had described.*

For both Bethe and Oppenheimer, the redemptive quality of love—*caritas*, charity[6]—became essential. Oppenheimer could voice, and perhaps only voice, "his deep yearning for friendship, for companionship, for the warmth and richness of human communication," for intimacy. "But neither circumstances nor at times the asperities of his own temperament permitted the gratification of this need in a measure remotely approaching its intensity."[7] He became a poet, the eloquent spokesman for meaning and for order in this new, infinite, chaotic world, trying through words to make the ideal of the infinite spiritual organism that Felix Adler had spoken of, "more and more potent and real in his own life and that of his fellow beings."

Bethe, on the other hand, has been able to express love through his work—by making his talents available to those for whom he cares and by creating a community. This has allowed him to find a balance between novelty and synthesis, and to be successful in the struggle to make partial order in the total chaos. He has also been able to learn to express his love openly in words. In his work and in his life, he indeed has helped himself, helped others, and helped mankind. It has not been an easy life, but he could accept the condition of man about which Oppenheimer had spoken. He could help, and he could find meaning, because he could come to love and be loved.

Notes to the Chapters

PREFACE

1. Ms. Velma Ray was Bethe's trusted and devoted secretary from 1948, when she first came to Cornell, until her death in the summer of 1992.
2. Schweber 1989.
3. See, for example, his fifty-page article on the theory of Type II supernovas that appeared in the *Reviews of Modern Physics* in 1993.
4. See Agar and Balmer 1998.
5. The *Bulletin of the Atomic Scientists* gives ample evidence of the scope of Bethe's activities in this realm and of his influence: there are frequent and extensive references to him, and it has published a large number of articles by him.
6. See, in particular, Stern 1969, Bacher 1972, and Bernstein 1982.

INTRODUCTION

1. Bainbridge 1975, 1976.
2. See, for example, Szasz 1984 and Hales 1997.
3. See chapter 1 of Smith 1965; the Franck Report is reprinted in Appendix B of that book. For a further account of the report, see Price 1995.
4. Smith 1965, 45.
5. Oppenheimer 1945, 7.
6. Ibid.
7. Price 1967.
8. Ibid.
9. Paul Forman, in an unpublished article, has insightfully examined the "Social Niche and Self-image of the American Physicists." He is particularly convincing in his analysis of the pre-World War II self-image of the American (experimental) physicist, which he shows combined "aggressive masculinity with self-sacrificing devotion to a transcendent, transfiguring cause" (Forman 1989).
10. Wigner 1947, v.
11. See Kevles 1995, 1990, Schweber 1988b, 1989, and particularly Forman 1987 and references therein for a discussion of the science-military partnership during the war and its dissolution thereafter.
12. See, for example, Rabi's recollections as recounted by Bernstein 1975 and Rigden 1987 and those of Wigner in volume 5 of his *Collected Works* (Wigner 1995).
13. See Rhodes 1986; Lanouette 1992.
14. Szilard 1978, 160.

15. See chapter 1 of Smith 1965 for some of the initiatives taken by Robert Wilson at Los Alamos and for an account of the activities at the Met Lab in Chicago.
16. Forman 1987; Schweber 1988b.
17. See Maila Walter's sensitive biography of Percy Bridgman, in particular the chapter entitled "Puritan Logic" (Walter 1990).
18. Bridgman 1947.
19. Ibid., 147.
20. Ibid., 150.
21. Ibid., 152.
22. Ibid., 154.
23. Dewey 1939.
24. Schweber 1988a.
25. Weber 1946, 156.
26. In an article reviewing physics in 1946, Philip Morrison (1947) characterized the year as one in which many plans, "often great and exciting ones," were being formulated, and he noted that the *Physical Review* was being spoken of as the *Physical PREview*. Morrison made particular reference to the large high-energy accelerators that were being built or had just gone into operation. He reported that as of November 1, 1946, the giant Berkeley cyclotron that Lawrence had built in 1940 and whose seventeen-foot magnet was used during the war as a prototype for the electromagnetic separation plant at Oak Ridge, Tennessee, had produced 200 Mev deuterons. At Illinois, Donald Kerst had begun the construction of a 300 Mev betatron, a 100 Mev version of which had successfully operated at the General Electric Laboratories in Schenectady, New York. At the University of California in Berkeley, William McMillan was building a 300 Mev electron synchrotron, as were teams at Michigan, MIT, and GE. Also at Berkeley, Louis Alvarez was engaged in building a large linear accelerator for protons, and it was expected that by February 1947 the first section delivering protons of more than 32 Mev would be operational. All these machines depended on the availability of new, high-power radar oscillators, a technology that had been developed during the war. In the second part of his article, in a section entitled "The Legacy of the War," Morrison reviewed some of the technical advances that the war had produced which would undoubtedly prove useful in experimental physics. He noted that during 1946 hundreds of war-experienced physicists had returned to their old laboratories "brimful of information about what had been done, and confident in their understanding of whole fields of technique which had been vague general possibilities in 1940." These new devices and techniques included microwave radiation ranging from a few millimeters to 30 cm, nuclear piles used as sources for both fast neutrons and well-collimated beams of thermal neutrons, and rockets for high-altitude cosmic-ray work.

The emphasis on the experimental practice in Morrison's article should be noted. Indeed, the dividends from the wartime activities manifested themselves there first.

27. The invited speakers included Isidor I. Rabi, who had returned to Columbia after serving as associate director of the Radiation Laboratory at MIT, and who addressed the issue of the support of research in universities by the government; Lee DuBridge, the recently appointed president of Cal Tech, who reflected on the problems stemming from the size of the laboratories that were being contemplated in nuclear research; the Yale philosopher F. S.C. Northrop, who presented his thoughts on the relation between the physical sciences, philosophy, and human values; Michael Polanyi, who grappled with the problem of how to justify the autonomy and freedom that the scientist demands and expounded his views on the foundations of freedom in science; and Percy Bridgman, who adumbrated new vistas for intelligence.

    See Sam Treiman's delightful and incisive introductory essay to the Festschrift celebrating Princeton's 250th Anniversary in which he comments on the presentations and discussions at the 1945 event in Fitch et al. 1997.

28. Wigner 1947, 31.

29. Harold Urey had won the Nobel Prize in 1934 for separating deuterium. During the war, he was in charge of a group investigating the feasibility of separating U^{235} from U^{238} by diffusion methods.

30. Wigner 1947, 40.

31. Kistiakowsky was a physical chemist, an expert on explosion. At Los Alamos he was in charge of the division that solved the problem of igniting a plutonium bomb using carefully timed focused implosions. He became President Eisenhower's science adviser. See Kistiakowsky 1976.

32. Wigner 1947, 40. There was also a heated discussion about Rabi's use of the word "parasitic" in his presentation. He had stated that since the aim of universities is to educate and to provide new knowledge, "university research is, to a certain degree, a part of the educational process and is parasitic on teaching" (Wigner 1947, 28).

33. Bush 1960, 18.

34. Wigner 1947, vi.

35. In the preamble to their report, the drafters of the Franck Report of June 1945 commented that "we found ourselves, by the force of events during the last five years, in the position of a small group of citizens cognizant of a grave danger for the safety of the country as well as for the future of all the other nations, of which the rest of mankind is unaware" (Smith 1965, 565). Similarly, in a speech to the Association of Los Alamos Scientists in early November 1945, Oppenheimer had stressed that "atomic weapons are a peril which affects everyone in the world, and in that sense a completely common problem" (Oppenheimer 1945).

36. Oppenheimer 1946, 9.

37. The title of Russell's lecture was "The Ivory Tower and the Ivory Gate." It is reprinted in Wigner 1947, 165–176.

38. Wigner 1947, 175.

39. Ibid. The distinguished British physiologist A. V. Hill had made a similar proposal earlier that year; see Hill 1946.

40. Smith 1965.
41. Oppenheimer had used these words in a discussion at which Bethe was present in Seattle during the summer of 1940.
42. To produce a chain reaction, one of the (on average) 2.5 (fast) neutrons released in the fission of a U^{235} nucleus must be slowed down to thermal energy and then captured by a U^{235} nucleus to induce a further fission. Fermi and collaborators constructed a "pile" of graphite blocks (to moderate the neutrons), which had a lattice of uranium metal inside. The pile was controlled by rods of boron, which are efficient neutron absorbers.
43. There exists by now a huge literature on the subject of the moral responsibility of the scientist, with different areas of science being addressed separately. See Shea and Sitter 1989; Erwin et al. 1994; Shrader-Frechette 1995. See also Proctor 1991 and Daston 1994. I found Iris Murdoch's 1992 *Metaphysics as a Guide to Morals* particularly valuable. Söderqvist 1996 has written incisively on the subject of the writing of biography and ethics.
44. I here use "parallel lives" not in Plutarch's sense. I am certainly not trying to deduce general traits or characteristic features of certain groups of human beings. Rather, I use "parallel" in the sense that their lives ran appositely, and tracing their intertwined lives highlights aspects of their lives that treating them individually would not. As will become clear in chapter 2, Bethe and Oppenheimer did indeed interact strongly during the course of their "parallel" lives.
45. For a discussion of the emergence of theoretical physics in Germany, see Jungnickel and McCormmach 1986; for the United States, see Schweber 1988a.
46. Weber 1946, 155.
47. Ibid., 142.
48. Teller 1962, 8.
49. Visvanathan 1988.
50. Douglas 1966.
51. I am here paraphrasing the remarks that Andre Lwoff made upon accepting his Nobel Prize in 1956: "Scientific research is a religion which calls for faith. Like every religion, it must have its prophets, a college of apostles and the heart and soul of a whole people. It must also have its martyrs."
52. Weber 1947, 329. Shils went on to say: "The central power has often, in the course of man's existence, been conceived as God, the ruling power or creator of the universe. . . . [However], the central power might be a fundamental principle, or principles, a law or laws governing the universe, the underlying and driving force of the universe. . . . Scientific discovery, ethical promulgation, artistic creativity, political and organizational authority . . . and in fact all forms of genius, in the original sense of the word as permeation by the 'spirit,' are as much instances of the category of charismatic things as in religious prophecy." See also S. N. Eisenstadt's introductory essay in Weber 1968.

53. The disparity between Oppenheimer's intensity and intellect and that of the other members of the community was readily apparent.
54. Weber 1947, 359.
55. Oppenheimer, in a speech on November 2, 1945, to the Association of Los Alamos Scientists, succinctly characterized this hope: "Atomic weapons constitute . . . a new field, and a new opportunity for realizing preconditions. I think when people talk of the fact that this is not only a great peril, but a great hope, this is what they should mean . . . [:] that in this field, because it is a threat, because it is a peril . . . there exists a possibility of realizing, of beginning to realize, those changes which are needed if there is to be any peace." Quoted in Rhodes 1986, 762.
56. Shils 1965.
57. For the distinction between pure and applied, and for an incisive overview of physics in the 1930s, see Weiner 1969.
58. Oppenheimer 1959, 19–20.
59. This stand resonated with the pragmatic philosophy of Peirce and James that Oppenheimer found congenial. Indeed, many of Oppenheimer's lectures after the war emphasized the role of scientific communities in a manner that has roots in Peirce's, James's, and Dewey's expositions. Furthermore, Oppenheimer did so well before Thomas Kuhn's *Structure of Scientific Revolutions* had made paradigms and scientific communities the nexus of scientific activities.

   The placing of morality into the human realm has of course a long history. Let me only quote Einstein: "My religious feeling is quite in line with Baruch Spinoza's pantheistic view. As a scientist I believe that nature is a perfect structure seen from the standpoint of reason and logical analysis. I believe, furthermore, that morality has nothing to do with religion in this highest sense but belongs in the human sphere entirely. But in the human sphere it is the most important thing." Einstein to Raymond Benenson, January 31, 1946, Einstein Papers, Albert Einstein Archives of the Jewish National and University Library, the Hebrew University of Jerusalem.
60. Bethe 1962.
61. They actually underestimated the critical mass by a factor of twenty. They also thought that it was possible to separate the two isotopes of uranium by thermal diffusion. The calculations were gone over, and the critical mass estimate was raised to about 3 kg in the MAUD report, which was made available to the United States in October 1941. The (somewhat optimistic) assessments in the MAUD report—regarding the amount of fissionable materials needed and the method of separation— were influential in Conant's and Bush's decision just before Pearl Harbor to approve a major American atomic bomb project. See "The Frisch-Peierls Memorandum" in Peierls 1997, 187–194.

   The initials MAUD do not form an acronym. They were given to the committee by G. P. Thomson, its chair, to mask its activities. Their origin is a cable in late 1942 from Lise Meitner to an English friend whose content was: MET NIELS AND MARGRETHE [BOHR] RECENTLY BOTH WELL BUT

UNHAPPY ABOUT EVENTS PLEASE INFORM COCKCROFT AND MAUD RAY KENT.
The message was passed on to John Cockcroft by Meitner's friend. Writing to James Chadwick, the head of the British atomic bomb project, Cockcroft decided that MAUD RAY KENT was an anagram for "radium taken," which agreed with other information available to the British that the Germans were getting hold of all the radium they could. Only later in 1943 did the committee learn that Maud Ray was the governess who had taught the Bohr children English and who lived in Kent.

62. Shils 1965.

63. Rabi depicted this "free but not anarchical" community as follows: "Members of this community possess an inner solidity which comes from a sense of achievement and an inner conviction that the advance of science is important and worthy of their greatest effort. This solidity comes in a context of fierce competition, strongly held conviction, and differing assessments as to the value of one achievement or another. Over and above all this too human confusion is the assurance that with further study will come order and beauty and a deeper understanding" (Rabi 1963, 308).

64. Oppenheimer 1946, 7.

65. Ibid.

66. Ibid.

67. Oppenheimer 1945 in Smith and Weiner 1980b.

68. Teller 1955, 274.

69. Feynman 1955, 1958.

70. Feynman 1993, 7–8. See also Feynman 1955, 13, and 1958, 261.

71. Rescher 1987, 9; italics mine.

72. The attempt to make the distinction between pure and applied physics can be seen in the separate departments of applied physics that were established at many American universities after the war. See Schweber 1988b, Kline 1996, Lowen 1997, and references therein.

73. In 1950, Oppenheimer told the high school students who had been finalists in the annual Westinghouse Science Talent Search that "in most scientific study, questions of good and evil or right or wrong play at most a minor and secondary part. For practical decisions of policy, they are basic" (Oppenheimer 1955, 109).

74. "Things like the atomic bomb, things like the great dam of the Colorado River, things like the jeep . . . I would like to call technology. They are impossible without science, and science in turn would be difficult and slow and impossible without technology. They are mutually fructifying. But they differ very much in their motivation, because surely the motive of any scientist, if he is an honest man, is that he wants to understand the world. And the motive of the engineer and the technologist is that he wants to change something about it and make it better. The one is aimed at understanding; the other at control" (Oppenheimer 1947, 9–10).

75. Oppenheimer 1955, 91.

76. See, for example, Richards 1978, Holton 1979, and Graham 1981. For a more recent discussion of these issues, see Rescher 1987 and Shattuck 1996.
77. Dyson 1997.
78. Boyer 1985; Weart 1988.
79. Oppenheimer 1947, 10.
80. Price 1967.
81. For a history of the GAC, see Sylves 1987.
82. Thus to point to two—among many—lectures: in the fall of 1947 Oppenheimer delivered the Arthur D. Little Lecture at MIT. Over one thousand people crowded into the auditorium to hear him. In the fall of 1948, Marshak invited him to talk in Rochester; over three thousand people came to listen.
83. See Smith 1965.
84. For a discussion of the issues involved, see B. Smith 1992 and references therein.
85. Thus, Oppenheimer initially supported the administration-backed May-Johnson bill that gave the military oversight of the development of atomic energy. Similarly, his statements at his trial regarding his former students may have been colored by his desire to retain his standing within the government.
86. See Hirschman 1970 for a fine study of the options available to individuals in various settings when confronted with policies with which they differ.
87. I characterize his role as new because in contrast to the scientist-statesmen of the 1930s—K.T. Compton, Conant, Bush—the scientist-statesman after World War II owed his influence and status to his expertise in current scientific and technical matters. This was the case for Oppenheimer, Szilard, Rabi, Bacher, von Neumann, among others.
88. In 1946 he wrote a paper with Bethe while visiting Cornell to deliver the Messenger lectures there. Bethe did all the calculations for the article (Bethe and Oppenheimer 1946). Another paper, published in 1948 with Lewis and Wouthuysen, was principally their work (Lewis 1947).
89. Lyndon Johnson to Prof. Hans Bethe, c/o Dr. James Perkins, President, Cornell University, October 16, 1966. Bethe Papers, Box 8, Folder 50, Cornell Archives, Ithaca, N.Y.
90. Nietzsche 1957, 7–8.
91. In his Reith lectures of 1953, Oppenheimer had asserted: "Our faith—our binding, quiet faith—is that knowledge is good and good in itself" (Oppenheimer 1989, 75).
92. Oppenheimer 1955, 121.
93. Quoted in Outram 1995, 1. But it should be noted that in 1956, Oppenheimer, after posing the questions: "Will this world, with its variety, its un-understood numbers, ever really yield to an ordered description, simple and necessary? Will our future students be able to explain the mass ratios, the coupling constants, the selection rules, as necessary consequences of the physical principles of the subnuclear world; or will

these remain empirical findings, to be measured with greater and greater accuracy, and recorded in tables that every physicist must memorize and carry with him?" gave the following answers: "Surely past experience, especially in relativity and atomic mechanics, has shown that at a new level of explanation some simple notions previously taken for granted as inevitable had to be abandoned as no longer applicable; but always in the past there has been an explanation of immense sweep and simplicity, and in it vast detail has been comprehended as necessary. Do we have the faith that this is inevitably true of man and nature? Do we even have the confidence that we shall have the wit to discover it? For some odd reason, the answer to both questions is yes" (Barton Papers, Box 119, Folder 3, Niels Bohr Library, AIP, College Park, Md.). Bethe would agree. His faith in reason has been constant.

94. I am here drawing on an unpublished lecture by Herbert Mehrtens delivered at the Dibner Institute at MIT in the spring of 1997. See also Weber 1946 and chapter 9 of Bauman 1997.

95. Fraternity is the appellation used by Oppenheimer (1947, 10). He repeated it in 1950. "The notion of a fraternity of scholars which is so familiar to Christian life . . ." (Oppenheimer 1955, 128).

96. Oppenheimer 1947, 18.

97. See Horkheimer and Adorno 1982.

98. See Sherwin 1987. See also Michaelis and Harvey 1973 and, in particular, Solomon 1973.

99. Weisskopf 1970, 29. In the fall of 1953, at the height of the McCarthy hysteria, the AAAS organized a symposium, "The Scientist in American Society," at its annual meeting. Victor Weisskopf was one of the speakers, and his paper was entitled "Science for Its Own Sake." He there characterized science as "the organized expression of the human trend to penetrate, clarify, and understand the world around us. But it is more. It has a universality and a validity independent of the individual's language in which it is expressed or created. Hence, it has a special human significance. It is a creation of mankind as a whole" (Weisskopf 1954). It should be noted that the rationalism, and associated insistence on the global relevance of science, that Weisskopf was extolling had been criticized after the war by Hans J. Morgenthau (1946) and Reinhold Niebuhr. Morgenthau believed that the scientists' internationalist bias, as reflected in the Baruch Plan that the United States had advanced for arms control, was symptomatic of an excessive faith "in the power of science to solve all problems and, more particularly, all political problems which confront man in the modern age" (Morgenthau 1946, vi). See Sims 1990, 60–64.

100. Oppenheimer 1945 in Smith and Weiner 1980b, 17.

101. "He came to my office," Truman wrote to Acheson in May 1946, "and spent most of his time ringing [sic] his hands and telling me they had blood on them because of the discovery of atomic energy." The meeting

between Truman and Oppenheimer probably took place on October 25, 1945. See Galison and Bernstein 1989, 278, and Stern 1969, 90.

102. Oppenheimer 1947, 10.

CHAPTER 1. WHAT IS ENLIGHTENMENT?

1. The essay is one of the last ones written by Foucault. It was published posthumously in Rabinow 1986. See Habermas 1985. See also Isaiah Berlin's essay "Kant as an Unfamiliar Source of Nationalism" in Berlin 1996, 232–248.
2. Foucault 1984, 32.
3. Foucault put it thus: It seemed to him that it was the first time "that a philosopher has connected . . . closely and from the inside, the significance of his work with respect to knowledge, a reflection on history, and a particular analysis of the specific moment at which he is writing and because of which he is writing. It is a reflection on 'today' as difference in history and as motive for a particular philosophical task that the novelty of this text appears" (Foucault 1984, 38).
4. Habermas 1986, 106.
5. Foucault 1984, 36.
6. Ibid., 37.
7. Ibid.
8. Ibid., 39.
9. Baudelaire 1964, 13.
10. Foucault 1984, 39–40.
11. Ibid., 42.
12. Ibid., 43.
13. Ibid., 47–48.
14. Ibid., 49; emphasis added.
15. Ibid., 50.
16. Both believed that of all intellectual activities, science—and perhaps only science—"has the kind of universality among men which the times require" (Oppenheimer 1947, 11). See also Oppenheimer's 1950 lecture, "The Encouragement of Science," in Oppenheimer 1955, 115.
17. Oppenheimer 1960, 13. In that same lecture, Oppenheimer noted that the fact that in all great cultures—the Christian, the Indian—"there is a notion that one does not return evil for evil, that one returns good for evil," and commented "what are we to think of our civilization and our society, which has lived for fifteen years now with the great new weapons of war, . . . with some ample public recognition of their scope and deadliness and what they could do, and which has never been able to come to grips with the questions of right and wrong, never found its way, never become oriented in the face of this new situation in any noble and any deep terms."

In a lecture entitled *Tradition and Discovery,* delivered during this same period, he was more explicit: "What is one to think of such a civilization which has not been able to talk about the prospect of killing almost everybody, or everybody, except in terms of calculation and prudence" (Oppenheimer 1984, 113). In his North Carolina lecture, he continued with the statement, "I have myself the feeling that to speak without the gravest of misgivings of the use of such means against enemies, who may indeed be very wicked enemies, in the way in which we have often done it, has been rather a disservice to the very cause of freedom in our country which they were intended to promote" (Oppenheimer 1960, 22–23).

18. Oppenheimer to George Kennan, October 4, 1951, Oppenheimer Papers, Box 43, George F. Kennan Correspondence, Manuscript Division, Library of Congress.

19. "There is for me, and I think there is for all of us, a very real mystery that so much order exists and that it is, in Jefferson's famous phrase, 'Fit for human understanding'" (Oppenheimer 1960, 18). See also Oppenheimer, "The Encouragement of Science," in Oppenheimer 1955, 114–115, and throughout his lectures collected in *The Open Mind*.

20. Howe 1990, 142.

21. Berlin 1980, 258. I would go so far as to suggest that Oppenheimer's willingness to have Los Alamos become a military base and he an army officer derives from his unresolved inner strife to find his "proper place." This unsettling ambiguity surfaced once again after the war in his support of the May-Johnson bill, which would have given the military control over atomic energy.

22. Berlin 1980, 258.

23. The brilliant mathematician Stanislaw Ulam was at Los Alamos during the war and returned to the laboratory in May 1946. He became deeply involved in the development of computing there and was responsible for the idea that made the American H-bomb possible. He got to know Oppenheimer well and had the following to say about him:

> Oppenheimer's opposition to the development of the H-bomb were [*sic*] not exclusively on moral, philosophical, or humanitarian grounds. I might say cynically that he struck me as someone who, having been instrumental in starting a revolution (and the advent of nuclear energy does merit this appellation), does not contemplate with pleasure still bigger revolutions to come. . . .
>
> It seems to me this was the tragedy of Oppenheimer. He was more intelligent, receptive, and brilliantly critical than deeply original. Also he was caught in his own web, a web not of politics but of phrasing. Perhaps he exaggerated his role when he saw himself as "Prince of Darkness, the destroyer of Universes." Johnny [von Neumann] used to say, "Some people profess guilt to claim credit for the sin." (Ulam 1987, 19)

24. In his postwar speeches he often noted that we live in an unusual world marked by very great and irreversible changes taking place on unprece-

dented timescales. For example, "We live in a time where our knowledge and understanding of the world of nature grows wider and deeper at an unparalleled rate; and where the problems of applying this knowledge to man's needs and hopes are new, and only a little illuminated by our past history" (Oppenheimer 1984, 123).

25. See, in particular, Oppenheimer's beautiful 1953 essay, "An Open House," in which he looks at the various rooms of the house called "science" (Oppenheimer 1984, 67–77). "It is a house so vast that none of us know it. . . . It is . . . arranged with a wonderful randomness suggestive of unending growth and improvisation. . . . [T]here are no locks; . . . no shut doors; wherever we go there are the signs and usually the words of welcome. It is an open house."

26. Oppenheimer 1955; "An Open House" in Oppenheimer 1984, 75–76.

27. Oppenheimer 1959, 39.

28. Oppenheimer, "Tradition and Discovery," a lecture delivered in 1960, in Oppenheimer 1984, 106. In another lecture from this same period, he asserted: "There are no synapses, no gaps, no gulfs in the picture of the various orders and parts of nature; but they are not derivable from one another. They are branches, . . . each with its own order, each with its own concern, each with its own vocabulary" (Oppenheimer 1960, 15). The metaphor of a tree is invoked because "it is a growing thing; it is a tree not a temple" (Oppenheimer 1960, 17).

29. Oppenheimer 1947.

30. Oppenheimer 1960, 12.

31. Oppenheimer 1980, 139.

32. Baudelaire 1964, 12.

33. See Camus 1962.

34. Nietzsche 1957, 6.

35. Jean Tatlock was the daughter of a professor of English literature at Berkeley and at the time was studying psychiatry at Stanford Medical School. "She was a beautiful and very intelligent woman who was fairly difficult to get along with and subject to fits of deep depression" (Serber 1998, 46). She and Oppenheimer had an intense on-again, off-again affair until the late 1930s. She committed suicide in January 1944.

36. See the letters he wrote to his brother concerning the appreciation of works of art and poetry, in particular the letter of December 30, 1928, in Oppenheimer 1980, 119–121. He then would probably have agreed with Baudelaire that "être un homme utile m'a paru toujours quelque chose de bien hideux" [to be a useful person has always seemed to me to be a most hideous thing].

37. Oppenheimer 1980, 135. One of Oppenheimer's friends indicated that "Robert could make grown men feel like school children. He could make giants feel like cockroaches" (Stern 1969, 126).

38. John Manley, the deputy director at Los Alamos during the war years, never observed the notorious Oppenheimer "putdown" at Los Alamos and was deeply impressed by Oppenheimer's control of his emotions and reactions. "[Oppenheimer] had no great reluctance about using

people," Manley told Alice Kimball Smith, "[but he did it] very adroitly. I think he realized that the other person knew that this was going on. . . . It was like a ballet . . . each one knowing the part and the role he's playing, and there wasn't any subterfuge in it" (Smith 1980b, 13). The "notorious" Oppenheimer putdown reemerged after the war. In a well-known incident, Oppenheimer in 1949 testified before the Joint Committee on Atomic Energy on whether radioisotopes should be allowed to be exported to friendly countries. The five commissioners earlier had voted 4–1 to permit such exports, Lewis Strauss casting the dissenting ballot. Strauss had objected to exporting such isotopes when they were to be used for industrial purposes. In his autobiography, Strauss later wrote that since the isotopes were to be used to develop alloys for jet engines operating at high temperatures, "that meant that the application must surely be for military purposes, since in 1949 jet engines were used only to power military aircrafts" and that this violated Section 10 of the Atomic Energy Act (Strauss 1962, 258). With Strauss present in the audience, Oppenheimer told the Committee: "No one can force me to say that you cannot use these isotopes for atomic energy. You can use a shovel for atomic energy; in fact, you do. You can use a bottle of beer for atomic energy. In fact, you do. But to get some perspective, the fact is that during the war and after the war these materials have played no significant part, and in my knowledge no part at all." And when then asked by Senator William Knowland whether it is not true that "the overall national defense of a country rests on more than secret military development alone," Oppenheimer replied: "Of course it does. . . . My own rating of the importance of isotopes in this broad sense is that they are far less important than electronic devices, but far more important than, let us say, vitamins, somewhere in between" (Stern 1969, 129–130).

39. Pais 1997, 239.
40. Oppenheimer 1980, 155–156.
41. We shall see that this metaphor had its origin in Felix Adler's writings on ethics—texts to which Oppenheimer was exposed at the Ethical Culture School.
42. See Herbert 1995, 149–150. I have quoted the version given in Bennett 1960, 178–79.
43. For Herbert, esthetics and theology were a seamless whole. Perhaps one of the attractions of Herbert's poetry for Oppenheimer was that he wanted to make esthetics and science a seamless whole. Isaak Walton, Herbert's biographer, suspected that Herbert had grave doubts about the theological value of his poetry. Oppenheimer, too, questioned the worth of his scientific production. Although Herbert did marry when he was thirty-six, only a few years before he died, he had decided at an early age "that the emotional peace and satisfaction he sought was not to be found in the love of women" (Bennett 1960, 59). Before he met Jean Tatlock, Oppenheimer probably had harbored similar thoughts.
44. Pais 1997, 241.
45. Clarke 1997.

46. There is an extensive literature on George Herbert. I found Bennett 1960, Roberts 1979, Vendler 1975, and Strier 1983 particularly helpful.
47. Hart in Roberts 1979, 454–455.
48. Recall Oppenheimer's statement to Truman about having blood on his hands when he and Acheson went to see him about the Acheson-Lilienthal Report, the first effort to achieve atomic weapons control through the ratification of an international atomic energy agreement.
49. Kennan first saw Oppenheimer in the fall of 1946 when Oppenheimer came to deliver a lecture at the National War College. In his *Memoirs*, Kennan described that encounter as follows:

> He shuffled diffidently and almost apologetically out to the podium: a frail, stooped figure in a heavy brown tweed suit with trousers that were baggy and too long, big feet that turned outward, and a small head and face that caused him, at times, to look strangely like a young student. He then proceeded to speak for nearly an hour, without the use of notes—but to speak with such startling lucidity and such scrupulous subtlety and precision of expression that when he had finished, no one dared to ask a question—everyone was sure that somehow or other he answered every possible point. [But] curiously enough, no one could remember exactly what he had said. The fascination exerted by his personality, the virtuosity of the performance, and the extreme subtlety of expression had actually interfered with the receptivity of the audience to the substance of what he was saying. This was to dog him throughout his life whenever it fell to him to address any other than a scientifically specialized audience. (Kennan 1972, 20)

50. Ibid., 18.
51. Ibid., 19.
52. Oppenheimer 1984, 138.
53. Baudelaire 1964, 8, 9.

CHAPTER 2. J. ROBERT OPPENHEIMER

1. From conversations with Oppenheimer, Rabi had the impression that his regard for the school "was not affectionate" and suggested that "too great a dose of ethical culture can often sour the budding intellectual who would prefer a more profound approach to human relations and man's place in the universe" (Oppenheimer 1967, 4). This was not the impression Oppenheimer gave in his interview with Thomas Kuhn for the Archive for the History of Quantum Physics, American Institute of Physics (AIP), College Park, Md. There, Oppenheimer expressed great admiration for the school's teachers and its academic program. There might have been some reasons for Oppenheimer's less than "affectionate" regard for the school—if indeed this was conveyed to Rabi—but

Rabi may have overstated the situation. Rabi, at times, had a tendency to forget certain events and to formulate his recollections to fit his assessments. See Forman 1998 for the contrast between Rabi's recollections of his college and early graduate education and the actual record.

2. Until the 1860s the sermons were delivered in German.
3. Later in life, Adler praised Hermann Cohen, who taught him philosophy, and more particularly, "the rigor, the sublimity, of Kant's system" (Adler 1918, 8–9).
4. Kraut notes that this "mission theory," which was formulated by the leaders of the German Reform Judaism movement, was important to German Jews because it furnished an ideology that gave them a raison d'être only in terms of religious identification, divested them of their nationality and nationhood, and thus made them more acceptable politically and socially in the German states. It was adopted as dogma by the American reform rabbinate. In both countries it served as a justification for keeping a separate Jewish identity and as a bulwark against assimilation (Kraut 1979, 24–25).
5. Friess 1981, 35.
6. Kraut 1979, 77–78.
7. See Birmingham 1967 for the story of the "great" Jewish families of New York, including the Seligman, Bache, Lewinsohn, Kuhn, Loeb, Lehman, Guggenheim, Goldman, Sachs, and Strauss families.
8. Kraut 1979.
9. For a thoughtful account of anti-Semitism in the United States, see the introduction by David Gerber of the volume he edited on the subject (Gerber 1986).
10. See Dinnerstein 1994, chapter 3.
11. Adler 1881.
12. Hermann Dühring and Heinrich von Treitschke were leaders in the movement, and Bismarck himself lent support to it in his attacks against the Liberal Party, prominent members of which were Jews.
13. Cohen 1978.
14. Adler 1881, 6.
15. See Birmingham 1967, 142–160; Dinnerstein 1994, 39–41.
16. For an account of the rise of the Seligman family, see Birmingham 1967.
17. Birmingham 1967, 147.
18. See Irving Howe's arresting narration of this history in his *World of Our Fathers* (1976). The immigration also exacerbated tensions and widened the chasm between the German and the other Ashkenazi Jews.
19. Dinnerstein 1994, 43–45.
20. Handlin 1959, 34.
21. For a narrative of the revitalization of the Reform movement, and an account of the Rabbinical Conference that led to the formulation of the eight-point "Pittsburgh Platform" redefining Reform Judaism's creed and practices, see Sarna 1998 and the references therein.
22. Kraut 1979.

23. See Kloppenberg's *Uncertain Victory* (1986) for a provocative and important study of Wilhelm Dilthey, Thomas Green, Henry Sidgwick, Alfred Fouillée, and William James, the writers he considers representative of the *via media* philosophers. The book also contains a valuable study of John Dewey.
24. Kloppenberg 1986, 28.
25. Lange published *Die Arbeiterfrage in ihrer Bedeutung für Gegenwart und Zukunft* in 1865, the same year the first edition of *Geschichte der Materialismus* was issued. An expanded second edition of this influential work appeared in 1873. Adler undoubtedly read Lange's *History of Materialism* and took many of the same positions as Lange had in that book. For some reason most of the analyses of Adler's philosophy have focused on the Kantian legacy, to the neglect of his readings of Fichte, Hegel, Lange, and the other German philosophers he studied while in Germany and thereafter.
26. Friess 1981, 35.
27. Adler 1918, 44–45.
28. Friess 1981, 51–52, 65.
29. Handlin 1959, 18–19.
30. The aims of the school were consonant with ideas put forward by John Dewey. The school would be "an embryonic community, active with types of occupation that reflect the life of the larger society. . . . For when the school introduces and trains each child of society into membership within such a little community, saturating him with the spirit of service, and providing him with the instruments of effective self-direction, we have the deepest and best guarantee of a larger society which is worthy, lovely and harmonious" (Dewey 1966, 15).
31. The competence and charisma of the teachers were indeed impressive. Oppenheimer raved about Augustus Klock, his physics teacher at the school. Henry Kelly was evidently as superb a teacher of biology. The caliber of the staff is exemplified by David Saville Muzzey, who was the head of the school until 1915, when he left to become the director of the Ethical Culture movement in St. Louis. A historian by training and avocation, he had written books on higher criticism and the prophets before joining the teaching staff at the Ethical Culture School (Muzzey 1900, 1902). In 1905 he wrote an article on the Franciscans which won him the Herbert Adams Prize of the American Historical Society (see Muzzey 1907). While teaching at the school, he wrote a two-volume history of the United States, which was extensively used, and a biography of Thomas Jefferson (Muzzey 1911, 1915, 1918). Both sets of books were widely acclaimed. He went on to earn a Ph.D. in physics at Harvard in 1931 with a dissertation on "Some Measurements of the Frequencies of Longitudinal Vibration for Bars of Magnetostrictive Materials."
32. The students also underwent regular medical examinations by a doctor attached to the school.

33. This was still true in the 1920s. See the report entitled *Education through Experience* by Mabel Goodlander (1922) of her four-year experiment at the school.
34. Adler 1892, 269–270.
35. Adler and Dewey were two of the most influential educational reformers during the period from 1880 to 1920. It should be noted that Adler's efforts started fifteen years before those of Dewey in Chicago. They held many similar views. Dewey (1899) advocated making schools "active with types of organizations that reflect the life of the society, and permeated throughout with the spirit of art, history and science." Dewey wanted schools to "saturate students with the spirit of service and to provide them with the instruments of effective self-direction." Adler and Dewey differed on the place of ethics in the curriculum. From the mid-1890s on, Adler was in contact with John Dewey. The two were summer residents at Glenmore, New York, and participated in the philosophical discussions that Thomas Davidson had organized there. From 1902 on, Adler held a part-time position as professor of social and political philosophy at Columbia and was thus a colleague of Dewey's, who had come to Columbia University and Teachers College in 1904.
36. Adler 1892, 258. In 1885 Adler began the first of three lectures entitled "Sketches of a Religion Based on Ethics" with the statement: "Our chief criticism of the ruling religious systems has been that they act too much on the emotions and too little on the will."
37. But Adler noted that "I do not say that rules are always good, but that the life of impulse is always bad" (Adler 1892, 48–49).
38. Friess 1981, 100.
39. The meeting place was on Central Park West at 64th Street; the school was a block away at 33 Central Park West.
40. Neumann 1916.
41. The school had expanded and moved to Fieldston in Riverdale in 1928.
42. Friess 1981, 136.
43. Ibid., 137.
44. Directory of the Society for Ethical Culture, 1917.
45. "[The teacher] is not to explain why we should do the right, but to make the young people entrusted to his charge see more clearly what is right, and to instill into them his own love and respect for the right. There is a body of moral truth upon which all good men, of whatever sect or opinion, are agreed: *it is the business of the public schools to deliver to their pupils this common fund of the of moral truth*" (Adler 1892, 15).
46. Ibid., 19.
47. See Lecture 20 in Dewey 1966, 200ff.
48. Dewey 1966, 200–201.
49. Adler 1892, 252.
50. Adler 1902, 180–181.
51. Adler 1902, 181.
52. Barbara Rosenkrantz stressed this last point to me.
53. Oppenheimer 1984, 137.

54. Directory of the Society for Ethical Culture, 1917, 9.
55. Friess 1981, 124–127.
56. See, for example, Bruford 1975.
57. See note 28, chapter 1.
58. Smith and Weiner 1980a, 6.
59. Thus, upon his arrival in Germany in the fall of 1926, he wrote Francis Fergusson: "Everyone . . . seems to be concerned with making Germany a practically successful & sane country. Neuroticism is very severely frowned upon. So are Jews, Prussians & French" (Smith and Weiner 1980a, 101).
60. In fact, Wyman recalled that Oppenheimer told him that music was positively painful to him (Oppenheimer 1980, 39).
61. Smith and Weiner 1980a, 91. See also Holton 1984.
62. Smith and Weiner 1980a, 92.
63. Ibid., 92–93.
64. Oppenheimer acknowledges the help of J. Edsahl [sic] in his paper, which was published by the Cambridge Philosophical Society in 1927.
65. Smith and Weiner 1980a, 94.
66. Holton 1984.
67. Schweber to Edsall, August 11, 1987.
68. I posed the same question to Alice Kimball Smith and asked her whether in her researches for her book with Charles Weiner she had come across any evidence that might support such a hypothesis. She answered that she "didn't have any *evidence* of sexual ambiguity to offer, which is not to say that it did not exist." It seemed to her that his ineptitude in the laboratory in addition to the normal growing pains of a sensitive individual were adequate explanation for the breakdown. A. K. Smith to S. S. Schweber, August 24, 1987.
69. Pais, in his autobiography, indicates that already in the early 1950s he "was convinced that a strong, latent homosexuality was an important ingredient in Robert's emotional makeup" (Pais 1997, 241). In Oppenheimer's FBI files, the following piece of "extremely derogatory information" was obtained during an interview with (name deleted) of the University of California at Berkeley: "It was further stated by (name deleted) that it was common knowledge on campus that prior to Oppenheimer's marriage he was possessed [sic!] with homosexual tendencies and at the time was having an affair with Harvey Hall . . . a mathematics student at the University, who was an individual of homosexual tendencies and at the time was living with Robert Oppenheimer." V. P. Keay to Mr. Ladd, November 10, 1947, J. R. Oppenheimer, FBI security file, microform, reel 1 (Wilmington, Del.: Scholarly Resources).
70. On his trip to New Mexico with Smith he met and was very attracted to Rosemary Horgan, Paul's sister, "a beautiful young woman, . . . dark-haired, delicate, and sensitive, who later developed a serious emotional illness." According to his brother, Frank, as a young man "Robert was more moved by her than by anyone" (Smith and Weiner 1980a, 40). She later married Alan Grant.

71. Oppenheimer 1980, 113.
72. There were evidently other women to whom Oppenheimer was close who similarly nurtured his political sensitivities, such as Estelle Caen and Sandra Dire Bennett. See Serber 1998, 31.
73. J. R. Oppenheimer, special radio program celebrating the sixtieth birthday of Albert Einstein, March 16, 1939. Oppenheimer's remarks were reported in *Science* 89, no. 2311 (1939): 335–336. In 1960 Oppenheimer attributed the spectacular success of the scientific tradition to "the methods and cooperative character of the undertaking . . . [which have] have proven to be golden, one of the great events of human history" (Oppenheimer 1960, 8).
74. In March 1934, Oppenheimer had responded favorably to an appeal to contribute between 2 and 4 percent of his annual salary to aid unemployed German physicists, but added: "I think the committee has been wise in not asking for help from American physicists." Oppenheimer to von Karman, in Oppenheimer 1980, 173.
75. Hilde Hein, personal communication.
76. For Oppenheimer's activities in the American Association of Scientific Workers (AASW), his connection to the Federation of Architects, Engineers, Chemists and Technicians (FAECT), and for the prewar history of AASW and FAECT, see Kuznick 1987, chapter 8.
77. Chevalier 1965, 33–34. In 1960 Oppenheimer took Einstein to task for equating his vision of laws, harmony, and regularity in the physical world with a vision of God. This seemed slightly "blasphemous" to Oppenheimer and a rather "contrived and inadequate testament to the true history of religion and its prophets" (Oppenheimer 1960, 18).
78. There was one fateful encounter between them in June 1943 when Oppenheimer visited her during a trip from Los Alamos to Berkeley. She was, at the time, undergoing psychiatric treatments. Jean Tatlock had been a Communist, and as the director of Los Alamos, Oppenheimer was under constant FBI surveillance. Their meeting became a focus of Oppenheimer's security trial in 1954.
79. Chevalier 1965, 30.
80. Katherine Peuning was born in Germany on August 8, 1910. She came to the United States with her parents in 1913. She married Joe Dallet, a trade union organizer in 1933 and became active in Communist circles. After his death in Spain, she attended UCLA and was married to Richard Harrison. For more on Joe Dallet, see Nelson 1981, Dyson 1984, 128–131, and the Oppenheimer FBI files, in particular 100–31936.
81. Smith and Weiner 1980a, 215.
82. Goodchild 1981, 128.
83. See Pais 1997, 242–243.
84. F. J. Dyson to S. S. Schweber, June 7, 1998.
85. George Eltenton was a British chemical engineer who had lived in the Soviet Union. In 1943 he was working for the Shell Development Corporation in Berkeley, and took an active interest in the recruitment efforts of the Federation of Architects, Engineers, Chemists and Technicians

(FAECT) at the Radiation Laboratory in Berkeley. He had met Oppenheimer, and Oppenheimer knew him to be an active member of FAECT.

86. Chevalier 1965, 65.

87. The Chevalier incident became one of the charges against him in his security trial.

88. Goldberg 1995, 483.

89. Berlin 1980, 258.

90. On August 7, 1945, upon hearing about Hiroshima, Chevalier wrote Oppenheimer: "There is a weight in such a venture which few men in history have had to bear. I know, that with your love of men, it is no light thing to have had a part, and a great part, in a diabolical contrivance for destroying them. But in the possibilities of death, are also the possibilities of life. You have made history."

91. Chevalier 1965, 70.

92. George Kennan, who knew him for sixteen years after the war, both at the State Department and at the Institute for Advanced Study, put it thus: "There was no one who more passionately desired to be useful in averting the catastrophes to which the development of the weapons of mass destruction threatened to lead. It was the interests of mankind that he had in mind here . . . the U.S. Government never had a servant more devoted at heart than this one, in the sense of wishing to make a positive contribution" (Kennan 1967, 55).

93. See, for example, the lectures he delivered between 1946 and 1953 on atomic energy and atomic weapons reprinted in Oppenheimer 1955. As is well known, he was the scientific member of the Lilienthal committee and helped draft the report that would have placed all atomic developments under an international agency responsible to the United Nations. A proposal embodying the outline of the Acheson-Lilienthal report—but not its spirit—was presented to the United Nations by Bernard Baruch, who placed emphasis on punishment for violations. Baruch in fact did not believe in arms control. The Russians who saw the "Baruch Plan" as a way to perpetuate the American monopoly in atomic energy and would not give up any of their sovereignty in favor of an international agency refused to consider the proposal.

94. In the late 1960s, Jerome Weisner commented that America's "virulent anticommunism will be seen as a mental disease."

95. Bethe 1968a, 390–416.

96. For a list of Oppenheimer's publications, see Oppenheimer 1967 and Bethe 1968a.

97. Smith and Weiner 1980a is the best introduction to the young Oppenheimer.

98. Interview with J. Robert Oppenheimer by Thomas S. Kuhn, November 18, 1963, 2. Archive for the History of Quantum Physics, AIP.

99. Harwood 1987, 1993, 1997.

100. Smith and Weiner 1980a, 77.

101. Kuhn interview, 4 and 9.

102. Smith and Weiner 1980a, 77.

103. Ibid.
104. Kuhn interview, 10.
105. I. I. Rabi, in Oppenheimer 1967, 3.
106. Oppenheimer 1926a,b,c. In papers b and c, he thanks Fowler and Dirac for their criticism, advice, and valuable suggestions.
107. E. U. Condon, "Autobiography. Notes," chapter 7, "Germany," 12. Condon Papers, American Philosophical Society (APS), Philadelphia.
108. Born and Oppenheimer 1927.
109. See Georgi 1989; Weinberg 1995–96.
110. Oppenheimer 1928; Gamow 1928; Gurney and Condon 1929. See Oppenheimer 1984, 173, for his recollection of these researches.
111. Serber 1967, 36.
112. Kuhn interview, 7.
113. Ibid., 8.
114. Ibid., 9.
115. Ibid., 10.
116. Oppenheimer 1930b.
117. Heisenberg and Pauli 1929, 1930.
118. Dirac 1928.
119. "The energy level of the normal state is thus infinitely displaced by the interaction of the particles with the field; the question which we have now to consider is whether or not the energy differences between two states are displaced by a finite or an infinite amount" (Oppenheimer 1930a).
120. The Lamb shift is the energy difference between the $2S_{1/2}$ and $2P_{1/2}$ state in hydrogen. The Dirac equation for an electron in a Coulomb field predicts that these two states have the same energy. But in 1947 Willis Lamb established experimentally in a precise and reliable manner that these levels are not degenerate but were separated by an energy difference of some 1050 Megacycles. See Schweber 1995.
121. Dirac 1930.
122. Pauli's proof is reported in a letter from Igor Tamm to Dirac, November 11, 1930. Dirac Papers, Florida State University, Dirac 3/3.
123. Oppenheimer 1930b.
124. Dirac 1978, 20.
125. Oppenheimer 1930b.
126. Tamm 1930.
127. Dirac 1931, 61.
128. In Bristol, Dirac went on to say: "There being no holes we can call protons, we must assume that protons are independent particles. The proton will now also have negative energy states, which we must again assume to be all occupied. The independence of the electron and proton according to this view allow us to give them any masses we please, and further there will be no mutual annihilation of electron and protons." But the difficulties had been gotten over "only at the expense of the unitary theory of the nature of electrons and protons." However, Dirac was not yet ready to give up his unitary view, for he added: "At the pres-

ent it seems too early to decide what the ultimate theory of the proton will be. One would like, if possible, to preserve the connection between the protons and electrons, in spite of the difficulties it leads to." By 1931, in his paper on "Quantized Singularities in the Electromagnetic Field," Dirac was ready to accept the independence of electrons and protons:

> It thus appears that we must abandon the identification of the holes with protons and must find some other interpretation for them. Following Oppenheimer, we can assume that in the world as we know it, all, and not nearly all, of the negative energy states are filled. A hole, if there were one, would be a new kind of particle, unknown to experimental physics, having the same mass and opposite charge to an electron. We should not expect to find any of them in nature, on account of their rapid rate of recombination with electrons, but if they could be produced experimentally in high vacuum they would be quite stable and amenable to observation." (Dirac 1931)

129. Dirac 1977, 21.
130. Bohr at first did not believe in Anderson's discovery, not even after Blackett had found clear evidence for electron-positron pair production in cloud chamber photographs of cosmic ray showers. When the evidence became more and more convincing, one of Bohr's final remarks was: "Even if all this turns out to be true, of one thing I am certain: that it has nothing to do with Dirac's theory of holes!" Pauli likewise objected to Dirac's new version of his hole theory. He wrote him in April 1933: "I do not believe in your perception of 'holes', even if the existence of the 'antielectron' is proved" (Pauli 1985, 159).
131. See, for example, his lecture in 1941 at the bicentennial conference of the University of Pennsylvania (Oppenheimer 1941). For a history of meson physics during the 1930s, see Galison 1987.
132. Serber 1967, 37.
133. This was the case in the 1931 calculation by Hall and Oppenheimer of the relativistic photoeffect.
134. Fowler was the senior theoretical physicist at Cambridge University. During the 1920s he had done important work in statistical mechanics and in astrophysics. He was Rutherford's son-in-law, a good friend of Bohr, and had been Dirac's teacher. See the entry on him in the *Dictionary of Scientific Biography.*
135. R. H. Fowler to E. C. Kemble, November 30, 1933. Kemble Papers, Harvard Archives, Cambridge, Mass.
136. R. Serber, "Particle Physics in the 1930's: A View from Berkeley," in Brown and Hoddeson 1983, 206. See also his interview with C. Weiner and G. Lubkin, 12, Center for History of Physics, AIP.
137. Bethe 1968a, 392.
138. W. H. Furry, interview with J. Ovgard and S. S. Schweber, June 27, 1979. Serber tells the following story: After obtaining his Ph.D. with John Van Vleck in Wisconsin in 1934, he had planned to go to Princeton to work

with Eugene Wigner, but on the way east, between Madison and Princeton, he stopped to attend the Summer School at Ann Arbor. Oppenheimer was there lecturing. He became "fascinated" with Oppenheimer and changed his fellowship plans and went to Berkeley. Interview of Serber with C. Weiner and J. Mehra, 7, Center for History of Physics, AIP.

139. Fermi 1932.
140. Breit 1932.
141. Bohr and Rosenfeld 1933.
142. See, for example, the Pauli correspondence with Bohr, Dirac, Einstein, Heisenberg, and others in Hermann et al. 1979–1981. The Weisskopf-Heisenberg correspondence is among the Weisskopf Papers, MIT Archives, Cambridge, Mass.; one-half of the Dirac-Bohr correspondence is in the Bohr Collected Works (1972–), and the other half is in the Dirac Papers in the Archives of Florida State University. Similarly, the Peierls-Dirac correspondence is in the Dirac Papers, Florida State University, and the Peierls Papers in the Bodleian Library, Oxford University. A. Kojevnikov edited the Dirac-Tamm correspondence.
143. Waller 1930.
144. W. Furry to V. Weisskopf, June 21, 1934, Weisskopf Papers, MIT Archives, Cambridge, Mass. A draft of this letter addressed to Dr. V. Weisskopf is in the Furry Papers, Harvard Archives, Cambridge, Mass.
145. Weisskopf 1934a.
146. W. H. Furry, interview with J. Ovgard and S. S. Schweber, 1979.
147. W. Furry to V. Weisskopf, June 21, 1934. Weisskopf Papers, MIT Archives.
148. Weisskopf 1934b. Weisskopf also wrote Furry thanking him for pointing out the mistake, and informing him of the final result; V. Weisskopf to W. H. Furry, undated, Furry Papers, Harvard Archives.

There is an interesting postscript to the story which is a testimony to the frailty of human memory. Weisskopf at some stage clearly forgot the content of the letter Furry had written to him in 1934. That letter explicitly stated that when they (Carlson and Furry) had read Weisskopf's paper, "The result for the electrostatic proper energy was new to us, as for some reason we had not previously realized the need for recalculating it." Many years later, Weisskopf characterized his handling of the letter as "one of the dark moments of my professional career." He recalled that "I made a mistake in the first publication that resulted in a quadratic divergence of the self-energy. Then I received a letter from Furry, who kindly pointed out my rather silly mistake and that actually the divergence is logarithmic. Instead of publishing the result himself, he allowed me to publish a correction quoting his intervention. Since then the discovery of the logarithmic divergence is wrongly ascribed to me instead of to Furry." V. Weisskopf, "Growing up with Field Theory: The development of QED," in Brown and Hoddeson 1983, 71. The record shows that Weisskopf was in fact the first to indicate how the self-energy must be computed in hole theory, and that *both* the electrostatic energy and the magnetic energy must be included in a hole-theoretic calculation. Had Furry and Carlson published their result in 1933, their calculation would have been incomplete. By not publishing their correction to the mistake in Weisskopf's paper, they acknowledged that Weisskopf had been the first to fully and correctly understand the self-energy problem in positron theory. When I pointed out to Weisskopf that Furry and Carlson had made only a partial calculation, Weisskopf replied to me: "But, still, it was a dark moment in my career when I made such a silly mistake in a most fundamental problem." V. Weisskopf to S. S. Schweber, August 24, 1985.

149. Bethe 1968a.
150. Serber 1967, 38.
151. Serber 1998, 42.
152. On arriving at Berkeley in 1934 as a National Research Council (NRC) postdoctoral fellow, Edwin Uehling made the following entry in his diary: "Thursday Aug 16 went to Le Conte Hall in the afternoon, found about a dozen physicists assembled in 219, the room assigned to me and apparently several others, joined in the festivities, and spent several hours doing things other than those for which I went there to do. Berkeley apparently a very sociable place." E. A. Uehling Papers, Diary 1934/5, University of Washington Archives, Seattle. See also Serber's account of the social life in his account of the "The Early Years" in Oppenheimer 1967.
153. According to Oppenheimer's FBI files, when his father died on September 20, 1937, he inherited $392,602, from which he obtained an annual dividend income of between $10,000 and $15,000, a considerable sum of money. His annual salary as a professor of physics was approximately $6,000.
154. Serber detailed the contribution of Oppenheimer and that of his students to "particle physics" during the 1930s in his "Particle Physics in the 1930s: A View from Berkeley" in Brown and Hoddeson 1983, 206–221.
155. Uehling 1935.
156. See Oppenheimer and Plesset 1933; Oppenheimer 1936; Carlson and Oppenheimer 1937. See also his other papers on these subjects listed in Smith and Weiner 1980a, 361.
157. See his papers with Melba Phillips and with Kalkar and Serber.
158. Davis 1968, 50.
159. Heilbron and Seidel 1989. The first thing that happened to Edwin Uehling upon his arrival at Berkeley in August 1934 as an NRC postdoctoral fellow, was to be taken through the Radiation Laboratory by Edward Nedelsky, another theoretical postdoctoral fellow. He made another visit shortly thereafter and entered into his diary that "the radiation laboratory [is] probably the most exciting laboratory in modern physics in existence." Uehling Papers, Diary 1933/4, entries for August 14 and August 27, 1934, University of Washington Archives.
160. Davis 1968, 50–51.
161. Robert Wilson, interview with Spencer Weart, May 19, 1977, Center for the History of Physics, AIP.
162. Davis 1968, 51.

163. Elsasser 1978, 199–200.
164. Serber 1998, 203.
165. The masslessness of photons implies that the range of electromagnetic forces is infinite. In Yukawa's theory, the exchanged quanta are massive, and the range of the resulting interaction, R, is related to the mass, μ, of the quanta by $R = h/\mu c$. This association of interactions with exchanges of quanta is a general feature of all quantum field theories.
166. They also determined its mass (150–220 electron masses) from measurements of the ionization it produced and from the curvature of its track in a magnetic field. Its lifetime was estimated to be about 10^{-6} sec. By 1939, Bethe could assert that "it was natural to identify these cosmic ray particles with the particles in Yukawa's theory of nuclear forces."
167. Tolman wrote one of the most thorough treatises on general relativity in 1934. In his preface he gratefully acknowledged discussions with Oppenheimer. See Tolman 1934. Zwicky had obtained his Ph.D. at the ETH in Zurich in 1922 with a dissertation under Peter Debye on the theory of ionic crystals. He went to Cal Tech in 1925 as a postdoctoral fellow, and remained there. In 1934, together with Walter Baade, he wrote two influential papers on supernovas and on neutron stars as the remnants of supernova explosions. Thereafter he started an important program of searching for extragalactic supernovas. See Karl Hufbauer's entry on Zwicky in the *Dictionary of Scientific Biography*.
168. Oppenheimer and Volkoff 1939.
169. Oppenheimer and Snyder 1939.
170. Such letters are to be found among the Oppenheimer Papers, Manuscript Division, Library of Congress, Washington, D.C.
171. See Oppenheimer 1970.

CHAPTER 3. HANS BETHE

1. Shortly after the war—due to disruptions and lack of coal—several Frankfurt high schools had to meet at the school, and classes were also held in the afternoon.
2. Guggenheim was born in Frankfurt in 1906 and is a few months older than Bethe. His father was a merchant who was born in Worms and his mother came from Frankfurt. His family was traditional: they observed the Shabbat and kept Kasher. Guggenheim's first six years of schooling were in a Jewish day school. He thereafter went to the Goethe Gymnasium. Guggenheim's circle of friends were Jewish and Zionists. As a teenager he belonged to Blauweiss and Kadimah, two Zionist youth organizations. He did not socialize with Hans Bethe. Guggenheim's closest friend was Nahum Glatzer, who introduced him to the circle around Franz Rosenzweig, Martin Buber, and Nobel. (On the latter, see the last volume of the Leo Baeck Institute publications.) It was Nobel who advised him to go into medicine, which he did. In the fall of 1924 Guggenheim enrolled in University of Frankfurt. Albrecht Bethe was his teacher

in a 1925 or 1926 physiology class he enrolled in. He remembers him as a tall, attractive person. "A fascinating lecturer, always interesting and lively." Guggenheim contracted tuberculosis in 1929 and was advised by his physicians not to pursue a career in clinical medicine. He first did research in bacteriology in Berlin, and after he emigrated to Palestine, he did research in nutrition. He established the Department of Nutrition at Hebrew University in Jerusalem and was the professor of nutritional sciences at Hadassah Medical School. He retired in 1974, and since then has written two books on the history of nutritional medicine. Interview with Karl Guggenheim in Jerusalem, August 17, 1993.

3. Goldschmidt changed his name to Ben Yosef and later joined a Kibbutz.

4. For the meaning of *Kulturträger* and what the education of one entailed, see Ringer 1969.

5. Mosse 1985, 3.

6. Sweet 1978, 12.

7. Friedrich Nicolai was the founder of the Allgemeine deutsche Bibliothek and owner of the largest bookstore in Berlin. He was a friend of Moses Mendelssohn and Gottfried Ephraim Lessing, whose *Nathan the Wise* was a bold attack on the prevailing prejudice against the Jewish minority.

8. See Altmann 1973. Markus Herz was a close friend of Kant. He had met Kant while studying in Königsberg in the late 1760s, and their correspondence is a record of the genesis of Kant's *Critique of Pure Reason*.

9. Sweet 1978, 18.

10. Herz 1988.

11. Sweet 1978, 39.

12. Humboldt 1975, 72.

13. Burrow 1975, Introduction.

14. Mosse 1985, 6–7.

15. Von Humboldt, quoted in Harwood 1993, 277.

16. One of the reasons that Humboldt could conceive of being able to implement such reforms at the university level was that the German universities enjoyed a semiautonomous corporate status. MaClelland (1980) has likened them to the medieval craft guilds. *Lehr und Lernfreiheit*, the freedom to teach and to learn, became the twin pillars on which neohumanism were anchored in the universities. They highlighted the autonomy of the academic disciplines.

17. Ringer 1969.

18. That Humboldt's commitment to Jewish emancipation was genuine is indicated by some of the actions he took. In 1814 he had written to the minister of justice to protest the fact that Jews were not permitted to perform autopsies and urged him to abolish such a measure based on prejudice. Humboldt was influential in charting the Prussian position regarding the rights of Jews at the Congresses of Vienna in 1814 and 1815. On June 14, 1815, Humboldt wrote his wife from Vienna shortly before the Congress adjourned: "I have always been favorably disposed towards this matter [the acquisition of equal rights by the Jews of Germany]. It is . . . an idea of my youth, for Alexander and I were regarded,

even when we were children, as bulwarks of Judaism." But his contact with Jews had greatly diminished after 1810—evidently by design. In another letter to his wife from Vienna, he indicated that his efforts there on behalf of the political rights of Jews were the last embers of his devotion to Henrietta Herz. "She has herself almost become a Christian. All are deserting the ancient gods."

19. Humboldt to the chancellor of Prussia September 30, 1816 in Kohler 1918, 36. Humboldt's paper is in Kohler 1918, 72–83.

20. It should be noted that Jews always constituted a small minority in Germany. In 1815 there were approximately 350,000 Jews in all the German-speaking states. Between 1871 and 1933, while the population of Germany increased from 40 million to over 65 millions, the size of its Jewish community remained fairly constant, numbering a little over half a million.

21. Sorkin 1983b.

22. Mosse 1985.

23. Ibid., 11.

24. Ringer 1969.

25. Mosse 1985, 4.

26. Needless to say, Jews desired the social standing that was attendant to being a high civil servant or university professor. These positions were perhaps the easiest avenue for social acceptance and complete assimilation.

27. On the other hand, Jewish students constituted approximately 7 percent of the total number attending Gymnasiums. At the time, Jews made up less than 1 percent of the total population of Germany.

28. Personal memoirs and other documents attest to the pressures exerted upon Jewish candidates to convert. By the early twentieth century, forty-four holders of university chairs were converted Jews and only twenty-five were Jews. The latter amounted to 2.5 percent of all the German *Ordinarii*—more than the percentage in the population but considerably less than the percentage of Jews in the student population. At this same time, Jews constituted 12 percent of the *Extraordinarii* and 14.5 percent of the *Privatdozenten.*

29. See Pauly 1987.

30. Harwood 1993.

31. See Phillipson 1962.

32. Jacobi was the son of a Potsdam banker. He obtained his Ph.D. in Berlin in 1825 and his Privatdozentship from Königsberg in 1826. He was appointed "extraordinary" professor in 1827 and ordinary professor in 1831. His enlightenment views are encapsulated in a famous letter he wrote to Dirichlet in 1830. "It is true that M. Fourier held the opinion that the principal aim of mathematics is public utility and the explication of natural phenomena; but a philosopher like him should have known that the only end of science is the honor of the human mind and that, in this respect, a question about numbers is worth as much as a

question about the system of the world." Jacobi to Dirichlet, July 2, 1830. The letter is quoted in Klein 1979.

33. Pyensen 1983; Jungnickel and McCormmach 1986.
34. In his obituary of Klein, W. H. Young noted that "Klein's two principles in recommending the call of a new professor were, first, individual pre-eminence, and second, collective representativeness (Young 1928).
35. Bethe 1933; Sommerfeld and Bethe 1933.
36. I. I. Rabi to F. W. Loomis, April 12, 1937, Pegram Papers, Box 89, Folder: Correspondence Rabi, 1936–1939. Columbia University Rare Book and Manuscript Library, New York.
37. Gleick 1992.
38. See Levi and Regge, 1989. Thus, both Einstein and Heisenberg, who were thought of as "geniuses" of the first kind during their astoundingly creative period when young, got mired in the formalism of the unified field theory they proposed as mature scientists and were never able to extricate themselves from the attraction of their theory.
39. See Traweek 1992 for a discussion of narrative strategies among physicists.
40. Chandrasekhar, Gamow, and Tuve 1938.
41. Bethe 1968b.
42. The diagnosis was bronchial TB. It is worth noting that thereafter Bethe was free of major illness throughout most of his life.
43. Bethe has never viewed himself as a Jew. He of course became aware of his connection to Judaism after the Nazis came to power. But his relationship to Jewry has never been close, neither before 1933 nor thereafter. He did not consider himself a Jew when he lost his position in Tübingen in 1933, nor did he let himself be made into a Jew by Hitler.
44. Sanders 1972.
45. Krieger 1992.
46. Ringer 1969.
47. See the list of contributors to the Festschrift in honor of Sommerfeld's sixtieth birthday in Debye 1928.
48. Sommerfeld 1928.
49. On the occasion of the eightieth birthday of his friend Victor Weisskopf, a symposium in his honor was held on October 12, 1988, at the American Academy of Arts and Sciences at which Bethe spoke. Bethe was introduced by Kurt Gottfried, who narrated the following story about him. In 1934 Weisskopf came to Bethe to tell him that he was about to undertake a calculation of pair production for spin zero particles which was similar to one that Bethe had performed the previous year for spin particles. Weisskopf wanted to know how long it would take to do the computation. Bethe answered: "Me it would take three days; you three weeks." At the start of his talk Bethe commented: "I was very conceited at that time. I still am—but I can hide it better."
50. Sommerfeld was offered the chair in theoretical physics in Vienna in 1916 and was invited to Berlin in 1927 to succeed Planck.

51. See the Festschrift in honor of Bethe's sixtieth birthday, edited by Marshak 1968.
52. This word is identified with Nazi rhetoric, so there is a touch of sarcasm in Bethe using it. It might be rendered in English as "national comrades."
53. Evidently, Sommerfeld wrote Bethe a personal letter inviting him to assume his chair, since Bethe's letter to Sommerfeld begins with the statement: "I have put off this letter for a long time because I was always hoping to get your letter. However, so far it has not arrived, so it probably got lost. But I know the essential content of your letter from indirect reports (from the FIAT, from my father-in-law and from Prof. Debye)" (Bethe to Sommerfeld, May 20, 1947, Sommerfeld Papers, Deutsches Museum, Munich).
54. Bethe to Oppenheimer, May 12, 1953, Oppenheimer Papers, Manuscript Division, Library of Congress.
55. If Beck, Bethe, and Riezler were somewhat contrite, George Gamow, who was in Copenhagen at the time, felt that pranks were his province and that intrusions could not be allowed to pass by unnoticed. Max Delbrück, who roomed together with Gamow at the Pension Have, recalled that on the day after Berliner's "Correction" appeared, Gamow had his plan. He would wait until another seemingly outrageous paper appeared in the *Naturwissenschaften*, and he would write to Berliner to persuade him that he had again been victimized by a prankster and pressure him to publish another retraction. Gamow did not have to wait long. On April 4 there appeared a note by A. V. Das on "The Origins of Cosmic Penetrating Radiation," which seemed to fill Gamow's bill. He got Rosenfeld and Pauli to agree that each one of them would write to Berliner to deplore the morals of the young generation, communicate their outrage concerning the Beck-Bethe-Riezler scandal, express satisfaction at seeing the Berliner correction, and profess their dismay at seeing Berliner victimized again. Gamow and Rosenfeld did write to Berliner. Pauli, who evidently had agreed to write his letter during a dinner at which a fair number of bottles of wine had been consumed, got cold feet in a more sober mood.
56. Sommerfeld and Bethe 1933; Bethe 1933.
57. Bethe 1986.
58. As for each of us, there are many parallel developments when following the trajectory of Bethe's life. On the personal side, as time went on, more and more bonds were made with other human beings, ties that connected him to lifelong friends, to his wife, to his family, to Cornell, to Los Alamos, i.e., to the many, often overlapping collectivities of which he is a member. Of importance is the fact that with the passing of time he was able to express more openly how meaningful these bonds were to him.
59. Bethe 1958.
60. It is important to note that the deep differences that Bethe has had over the years with Edward Teller have often stemmed from their differing

conception of what commitments were entailed by membership in the scientific community. Thus, in their disagreement over the development of the H-bomb in the fall of 1949, Bethe blamed Teller "for leading Los Alamos, and indeed the whole country, into an adventurous program on the basis of calculations that he must have known to have been very incomplete" (York 1989, 172).

61. Oppenheimer 1955, 94.
62. Ibid., 114–116.
63. I. I. Rabi, September 26, 1950, opening exercises, Columbia University, New York. In Oppenheimer Papers, Rabi Manuscript Division, Library of Congress.
64. See for example, Badash et al. 1980.
65. Oppenheimer 1970, 14.
66. Oppenheimer 1989, 71.
67. Bethe noted: "It was a very open and informal laboratory in which everybody talked to everybody, as long as they had a white badge. You couldn't talk to the people with red badges; they were service personnel." Everyone who had been cleared to work within the site—librarians, people carrying out the computations—had white badges.
68. Edward Teller, no friend of Oppenheimer, had the following to say about Oppenheimer's wartime leadership:

> Much of my life has been spent in laboratories of similar size and nature [as Los Alamos]. I have known many of the directors intimately. For a short time, I was even a director myself. I know of no one whose work begins to compare in excellence with that of Oppenheimer's.
>
> Throughout the war years, Oppie knew in detail what was going on in every part of the Laboratory. He was incredibly quick and perceptive in analyzing human as well as technical problems. . . . He knew how to organize, cajole, humor, soothe feelings—how to lead powerfully without seeming to do so. He was an examplar of dedication, a hero who never lost his humanness. Disappointing him somehow carried with it a sense of wrongdoing. Los Alamos' amazing success grew out of the brilliance, enthusiasm and charisma with which Oppenheimer led it. (Teller 1983, 191–192)

69. Oppenheimer Papers, Manuscript Division, Library of Congress.
70. In 1992 Bethe stated his views as follows:

> Los Alamos was Oppenheimer's time of glory, and nobody else could have done it. I am quite convinced of that. He had to do so many things at the same time. He had to keep this big group of prima donnas in order . . . the people in Bacher's division who were the top of the profession, several in my division, Kistiakowsky, then a man of completely different background and interest, Parsons . . . to keep all of us in line, to make all of our

work mesh, to understand it all. . . . He really understood every-thing. He was not productive like von Neumann, or Fermi. He had a number of good ideas, but in that respect he was a peer, and no more. But in knowing everything, he was unique. Know-ing everything and knowing where more work, where more ef-fort, was needed. And at the same time, he had to deal with Groves. . . . Groves was a very disagreeable man. And Groves was apt to consider all the scientists his slaves. Even if we were not in uniform, we were just lieutenants. And somehow Oppie managed to hold his own with Groves; to have Groves acknowl-edge that he, Oppie, knew what to do technically. And at the same time he was having the trouble with security. They were in-vestigating him, they followed him to his former girl friend, Jean Tatlock; they investigated his communist connections, and the unfortunate matter with Haakon Chevalier occurred then. The security people at Los Alamos gave him a hard time very often. At the same time there was the imposition of security at Los Alamos. He had to apologize, and make us accept the censor-ship. Then there were problems in town; there was a town coun-cil, of which Weisskopf was the chairman. And there were the close, and at times strained, relations between the army and the scientists. (Interview with S. S. Schweber, July 1992)

71. Oppenheimer 1970, 325.
72. Bethe recalled that not all of Oppenheimer's students were of a similar view. He remembers Hartland Snyder carrying a placard in Seattle in support of isolationism stating "The Yanks are not coming."
73. Interestingly, later that summer when Bethe was in Stanford for a month, he visited Oppenheimer once or twice in Berkeley, but only talked physics with him; they did not talk about politics. "It was peculiar that we did not talk about politics since we felt strongly and similarly about the war."
74. Joseph Weinberg came to Berkeley in February 1939 as a graduate stu-dent to work with Oppenheimer. He remembers that the graduate stu-dents at that time spoke of Oppenheimer as being very sympathetic to "left" causes, but that he was not thought to be a member of the Com-munist party, though he supported the party's aims. He was evidently a member of a "study group," some participants of which were mem-bers of the party. This was before the Molotov–von Ribbentrop Treaty was signed in June 1939. Telephone interview with Joseph Weinberg, November 13, 1997.
75. Oppenheimer had picked up the nickname "Oppje" (sometimes spelled Opje—which is the way Oppenheimer signed letters to his friends) dur-ing his stay in Leyden. All the letters from his friend George Uhlenbeck address him in this manner. By 1936 the appellation had become Ameri-canized to "Oppie." After his marriage to Kitty in 1940, she insisted he be called Robert.

76. F. Bloch to I. I. Rabi, November 2, 1938, Bloch Papers, General Correspondence, Box 1, Stanford University Library, Stanford, Calif. Rabi had been a lecturer at Stanford during the summer session of 1938.
77. Dewey 1937a.
78. Dewey 1937b.
79. Ibid.
80. Oppenheimer 1980, 196.
81. See Gowing 1974, Rhodes 1986, and Peierls 1997.
82. Bethe recalls that the group had available the Tube Alloys reports that covered the work the British had done up to that point, in particular the paper that Dirac had written in which he estimated "the efficiency of energy release with a non-scattering container." See Dalitz in Kursunoglu and Wigner 1987, 69–92.
83. Teller 1955, 269.
84. "There are two men whom I should be more than reluctant not to have on the project: Bethe and Bacher. I think that you know the reasons in each case, and agree with them. You have a great deal of influence with these two men, and they in turn on many others who are involved in the project. I am asking that you use that influence to persuade them to come rather than stay away." Oppenheimer to Rabi, February 26, 1943, Oppenheimer Papers, Box 6, Manuscript Division, Library of Congress. See also Oppenheimer's letter to Rose and Hans in Oppenheimer 1980, 242–244.
85. Details of the invitation can be found in Oppenheimer's FBI file. Oppenheimer gave six lectures during the first two weeks of May 1946. The titles of the talks were "The Present Crisis: Atomic Weapons"; "Atomic Energy"; "The Nature of Change in Physics: (1)Complementarity, (2)The Atomic Nucleus, (3)Elementary Particles"; "The International Control of Atomic Energy." Each had an attendence of between 1,000 and 1,200 people, consisting almost entirely of Cornell students and faculty members. A "confidential informant T-1" relayed to the FBI the amount Oppenheimer was paid ($1,500), when he came and where he lived in Ithaca, whom he saw while there, and how he spent his time. Report by D. E. Roiey, AI File 100–9401, dated 8/5/46, J. R. Oppenheimer, FBI Security File, microform, reel 1 (Wilmington, Del.: Scholarly Resources.) In subsequent reports it was inadvertently revealed that the "confidential informant T-1" was Edward Graham, the secretary of Cornell University!
86. Remarks at department meeting, April 23, 1947, Oppenheimer Papers, Manuscript Division, Box 230, Library of Congress. Also, Birge Correspondence, 1943–47, University of California, Berkeley.
87. Robert Sproul, the president of the university, in his letter to Bethe inviting him to assume the chair of theoretical physics, wrote: "From your visit of last winter as Hitchcock Professor [you will know] how great are the opportunities and how great are the needs for a man of your preeminence." The salary Bethe was offered was $12,500 a year, which Sproul indicated was "higher than that of any professor at the University of California" and indicative of "the importance we attach to having you

with us." Oppenheimer Papers, Manuscript Division, Box 230, Library of Congress. Also Birge Correspondence, 1943–47, University of California, Berkeley.

88. See Schweber 1995, chapter 5.
89. Ibid.
90. Bethe was in fact chair of the committee that analyzed the monitored air samples; see Ziegler 1988.
91. See Rhodes 1995.
92. Oppenheimer's telegram is dated October 20, 1966, and is in Bethe Papers, Box 8, Folder 50, Cornell Archives. Bethe's note is in the Oppenheimer Papers, Archives Division, Library of Congress.
93. Peierls 1997, 55.
94. Bethe 1968a.

CHAPTER 4. THE CHALLENGE OF McCARTHYISM

1. See Jensen 1991.
2. See Rhodes 1995 for a full account. Also see "Rethinking McCarthyism, If Not McCarthy," sec. 4, p. 1, of the *New York Times*, October 18, 1998.
3. Schrecker 1994.
4. Un-American Activities Committees were set up in Congress and in various state legislatures to investigate subversive organizations, Fascist as well as Communist. For a history of HUAC until 1950, see Carr 1952. See also Wang 1992, 1998.
5. Carr 1950, 37–38. Thomas resigned from the House in January 1952 after being convicted in a federal court for fraud.
6. Nelson et al. 1981.
7. For a background of Steve Nelson, see his autobiography, Nelson et al. 1981. For his depiction as seen through the eyes of HUAC, see the foreword to the June 8, 1949, hearing of HUAC, v–ix, Congressional Record. The description makes a point to connect Nelson with Joe Dallet and with Katherine (Kitty) Puening Dallet Harrison, whom Oppenheimer had married. Kitty's former husband, Joseph Dallet, had been a member of the Communist party, had been a good friend of Steve Nelson, and had died fighting on the Loyalist side in Spain. Kitty herself had joined the party (see Stern 1969, 28–30). The letters of Joe Dallet to his parents, as well as other materials relating to his activities as a labor organizer during the 1930s, are in the Brandeis University Farber Archives, Waltham, Mass. The Brandeis Library contains a very extensive collection of materials pertaining to the Spanish Civil War, including materials on Steve Nelson. For further insights on Joe Dallet, see Dallet 1938 and Dyson 1984.
8. HUAC hearing, 81st Congress, 1st sess., foreword, v–vi. Scientist X was charged by HUAC to have given to Steve Nelson "late one night in March 1943 . . . a complicated formula" relating to the atomic bomb project. The HUAC hearing regarding Communist infiltration of the Radiation

Laboratory and Atomic Project at the University of California, Berkeley, took place during the first session of the 81st Congress on April 22 and 26, May 25, June 10 and 14, 1949. These hearings came on the heels of the committee's unsuccessful attempts the previous year to find any evidence of atomic spying.

9. Even though the hearings had been held in closed executive sessions, some of its proceedings were leaked to the press. Thus, on September 14, 1948, Condon wrote Bernard Peters: "As you probably know your name was in the Washington *Times-Herald* one day last week in a derogatory way in relation to the House Committee on Un-American Activities." Condon Papers, Condon-Peters correspondence, American Philosophical Society (APS) Archives, Philadelphia.

10. Bohm, Fox, Lomanitz, and Weinberg all had been active members of the Federation of Architects, Engineers, Chemists and Technicians (FAECT), a "left wing" union affiliated with the Congress of Industrial Organizations (CIO) during their wartime association with the Radiation Laboratory. Such unionizing activities had been forbidden within the Manhattan Project. Most likely, their problems stemmed from the fact that Oppenheimer had identified them as being left-wingers, and perhaps Communists, during his clearance proceedings. A *Rochester-Times Union* story dated June 15, 1949, stated that "Dr. Oppenheimer became 'acquainted' with the existence of a Communist cell at Berkeley by disclosure of the intelligence agencies of the Government."

11. Only on March 15, 1950, in the report of its activities during 1949, did HUAC identify Weinberg as Scientist X. Weinberg appeared three times before the committee and denied the charge of having given secret documents to Nelson. He also denied ever having known Nelson. The committee found that this last declaration contradicted the testimony of another witness, a former intelligence agent. HUAC recommended that Weinberg be prosecuted for perjury. In May 1952 a grand jury indicted him for perjury but he was eventually acquitted.

When he came to Berkeley in the spring of 1939, Joseph Weinberg was very much "left of center" politically, and strongly approved of the Soviet Union's support of the Loyalist cause in Spain, *but was never a member of the Communist party.* He didn't see very much of Oppenheimer socially, "perhaps once or twice a year at a Ph.D. party," and therefore does not know where Oppenheimer formed his impression of Weinberg's political views. Telephone interview with Joseph Weinberg, November 13, 1997.

12. In the summer of 1947, while still in Berkeley (before accepting a position at the University of Minnesota), Joseph Weinberg was interviewed by the FBI. Weinberg recalls that "they were mostly interested in JRO. They asked about Frank, Kitty. . . ." Weinberg blames Velde (then on the investigative staff of HUAC) "as the one who was after Oppenheimer, and thereafter after him." Telephone interview with Joseph Weinberg, November 13, 1997.

13. When Peters got back to Rochester, he was dissatisfied with the few questions the committee had asked him since it had subjected him to a large amount of derogatory publicity the year earlier. He wrote to Representative John Wood, the chairman of HUAC, asking for another hearing to go into the matters more fully. He did not receive an answer. Edward Condon to his wife June 23, 1949. The letter is reprinted in the *New York Daily Mirror*, April 16, 1954. The letter from Condon to his wife was intercepted and opened by the FBI. It became part of the file the FBI kept on Condon. It was referred to at Oppenheimer's trial, and Condon, upon reading the transcript, wrote a deposition (which he, however, never submitted) in which he outlined his interactions with Oppenheimer. Condon Papers, APS, Philadelphia.
14. Oppenheimer was also asked about Haakon Chevalier and his brother, Frank. See Stern 1969, 120–122.
15. Ibid., 118–119.
16. Hearings, Joint Committee on Atomic Energy, Investigation into the United States Atomic Energy Project, Monday, June 13, 1949. Oppenheimer was appearing before the committee in his capacity as chairman of the AEC General Advisory Committee to defend the GAC's recommendation to allow the shipment of radioisotopes to foreign countries, in particular the shipment of radioactive iron to the Norwegian Atomic Energy Commission. This was the affair being featured in the *New York Times* June 9 headline. It was on this occasion that Oppenheimer ridiculed Admiral Strauss, one of the AEC commissioners, who was opposed to the isotope shipment policy. Strauss never forgave him for this humiliation.
17. *Time* magazine, June 27, 1949. This is the *Time* coverage of Oppenheimer's appearance before the committee to defend the GAC's recommendation to allow the shipment of radioisotopes to foreign countries for medical and research purposes.
18. The family was Jewish; his name on his birth certificate was Bernhard Pietrkowski. His father was a physician who also did pharmacological research.
19. Lal 1993, 612. The story as related by Oppenheimer has Peters's mother securing his transfer from Dachau to a Munich jail, from where he escaped and in 1934 found his way to the United States.
20. Peters and Oppenheimer may have been distant relatives by marriage.
21. These notes became the basis for Schiff's book on quantum mechanics.
22. Interview with Philip Morrison, July 30, 1997.
23. Morrison 1991, 743.
24. When they first came to San Francisco, Hannah Peters became a close friend of Jean Tatlock, with whom Oppenheimer was romantically involved at the time. She occasionally also served as Oppenheimer's personal physician.
25. Oppenheimer 1970, 120–121.
26. Peters had become a U.S. citizen in 1940.
27. Oppenheimer 1970, 121.

28. The quotation is from Oppenheimer's testimony in 1954 before the Gray Board. He could not recall the fourth name.

29. Oppenheimer 1970, 150. When later this incident was brought up at his security trial in 1954, Oppenheimer disavowed "that this was something dredged up for [de Silva]," denied having characterized Bohm as "truly dangerous" since "he never thought of him that way," and believed that "there is a garble in this and also that the tone is not . . . accurate." But he did corroborate that he had stated that Peters was the most dangerous of the four names that de Silva had mentioned. Oppenheimer 1970, 150.

30. "The chemical composition of the primary radiation leads to rather interesting restrictions on the possible origin of cosmic rays." Peters to Condon, December 17, 1949. Condon Papers, APS. See also Morrison 1991.

31. The January 12, 1950, *New York Times* carried an article with the headline: "'Strong Evidence' of a New Atomic Particle, the Neutral Meson, is Hinted in Photographs." It reported that Helmut Bradt, Bernard Peters, and Morton F. Kaplon, all three of Rochester, were responsible for the discovery.

32. Oppenheimer 1970, 211.

33. Edward Condon to Emilie Condon, June 23, 1949. The letter was reprinted in the *New York Daily Mirror* on Friday, April 16, 1954.

34. There is no question that Oppenheimer would have read the June 9 story in the *New York Times* in which Wood, the chairman of HUAC, indicated that Peters had denied ever being a member of the Communist party. It is also likely that he read the Rochester *Times-Union* interview of Peters.

35. Weisskopf to Oppenheimer, June 27, 1949. Oppenheimer Papers, Box 77, Manuscript Division, Library of Congress.

36. In the draft to the letter, Weisskopf, instead of writing "from you," had written "and that scientist is you." Weisskopf Papers, MIT Archives, 92–43, Box 3, Folder 42.

37. Weisskopf to Oppenheimer, June 27, 1949, Oppenheimer Papers, National Archives, Washington, D.C.

38. In the draft of the letter, Weisskopf had a postpostscriptum concerning Frank Oppenheimer which he crossed out: "As much as I could spank Frank for his idiotic behavior 2 years ago, as much do I admire his present stand before the committee." Weisskopf Papers, MIT Archives, 92–43, Box 3, Folder 42.

39. Somehow Weisskopf had forgotten—or suppressed—what had happened in the Third Reich when the charge of teaching Jewish physics was brought against Heisenberg and others by Philip Lenard and Johannes Stark. See Hentschel and Hentschel 1996. For an explicit linkage of Oppenheimer and Galileo, see de Santillana 1968, 120–136.

40. Frank Oppenheimer, Robert's brother, was an experimentalist doing research on cosmic rays and in 1949 was an assistant professor of physics at the University of Minnesota. Together with his wife he had appeared before HUAC on June 14. The hearing started in the morning in execu-

tive session, but before the day was over the committee voted to turn the hearing into a public one and made the full testimony available for public scrutiny. Frank acknowledged that he and his wife, Jacquinette, had been members of the Communist party in the late 1930s, but that he had resigned, disillusioned, three and a half years later, before he had started to work at the Radiation Lab. In 1947, when the Washington *Times-Herald* had related these facts, Frank had called the account a "complete fabrication" and had similarly disavowed the story in a letter his lawyer sent to the chair of the physics department at Minnesota. A few days before testifying before HUAC, Frank had submitted a letter to the administration of the University of Minnesota offering to resign his post as assistant professor of physics. Contrary to his expectations, his resignation was promptly demanded when his testimony made the headlines of the evening newspapers on June 14.

41. Peters and Condon had been friends since the early 1940s. After the war, they worked together within the Federation of Atomic Scientists (FAS) to get civilian control over atomic energy. After HUAC started "investigating" both Condon and the Rad Lab, they kept each other apprized of developments. They remained good friends throughout Condon's life. See the correspondence between them in the Condon Papers, APS. In a very long letter written on April 28, 1954, to Henry Luce, the editor of *Time* (but never sent), protesting the false impressions about his relation to Oppenheimer contained in the April 26 issue of *Time*, Condon asserted "I have regarded it as my Christian and patriotic duty to help in such a way as I can to sustain the morale of various young scientists who have come under attack." The young scientists he was referring to were David Bohm, Frank Oppenheimer, and Bernard Peters. Condon continued: "I do not wish to be misunderstood as condoning any improper behavior on their part where it is proven, but where it is not proven, or where it represents an emotional and unjustified exaggeration of old charges such as reached its climax against Robert Oppenheimer himself, then I have felt it important to help them hold together and to defend themselves partly out of decent consideration for them as individuals and partly out of a feeling of distress at seeing our country being scientifically weakened when we have no scientific strength to spare." Condon to Henry Luce, April 28, 1954, Condon Papers, APS. For more on Condon and his plight and stand during the McCarthy era and thereafter, see Wang 1992.

42. Condon to Henry Luce, April 28, 1954, Condon Papers, APS.

43. Condon to Oppenheimer, June 29, 1949, Oppenheimer Papers, Box 72, Manuscript Division, Library of Congress. See also Condon's testimony before HUAC on September 5, 1952, 3860–3865; and Oppenheimer 1970, 212. In a letter written a few days earlier to his wife, Condon had declared, "I am convinced that Robert Oppenheimer is losing his mind. . . . I understand that Oppie has been in a very high state of nervous tension in the last few weeks. People from Princeton say that he seems to be in a state of strain for fear he will be attacked. . . . It appears he is trying

to buy personal immunity from attack by turning informer, including certain imaginative fictional 'information' against his close personal friends." *New York Daily Mirror,* April 16, 1954.

44. Bethe to Oppenheimer. June 26, 1949, Oppenheimer Papers, Box 20, Folder: Bethe, Manuscript Division Library of Congress.

45. Ibid. In an attachment to his letter, Bethe sent Oppenheimer a copy of the letter "in behalf of Frank Oppenheimer" that had been sent to William Buchta, the chair of the physics department at Minnesota, where Frank held the position of assistant professor. It had been signed by many individuals at the conference. Bethe was careful to add that, "of course, these were actions of individuals and not of the conference. Millikan and Teller wrote two additional private letters to Buchta." He concluded his note by saying that it was good to have had Frank at the conference, "he had many interesting things to say (on physics)."

46. Oppenheimer 1970, 212.

47. Peters to Weisskopf, Weisskopf Papers, 92–93, Box 3, Folder 42, MIT Archives. Peters concluded his letter with: "If you get by Rochester this fall I would like to discuss with you some rather surprising types of nuclear collisions we have found recently."

48. Oppenheimer 1970, 214.

49. Ibid.

50. Condon to Luce, April 28, 1954, Condon Papers, APS.

51. Samuel Goudsmit looked into the matter and discovered that Oppenheimer's testimony to de Silva was part of Peters's file, and that this "derogatory information" was the source of the trouble Peters was experiencing. Goudsmit talked to Oppenheimer at the April 1949 meeting of the American Physical Society in Washington about the harm his statements were causing to his former student. According to Stern, Goudsmit asked Oppenheimer whether he still held the same opinion of Peters as he did five years ago and that Oppenheimer had answered: "Just look at him. Can't you tell that he can't be trusted?" Stern 1969, 122–23.

52. Peters to Condon, September 16, 1955. Condon Papers, APS.

53. A special issue of *Current Science* (vol. 61, no. 1, 1991) was published in honor of Peters's eightieth birthday. It contains several articles by colleagues and students outlining his varied contributions. Also included is an article by Peters of his reminiscences of his association with the Tata Institute of Fundamental Research.

54. The Institute is located in the Ionosphere Laboratory of Denmark's Technical University Lyngby.

55. Lal 1993.

56. Stern 1969, 214–218.

57. Ezra Day had resigned as president of Cornell in 1948 after having suffered a severe heart attack.

58. Wright to Morrison, April 5, 1951. Wright Papers, Cornell Archives, Ithaca, N.Y.

59. Wright quoted these passages in the letter he sent to Morrison on the day of his meeting with him. The emphasis was added by Wright.

60. Wright to Morrison, April 5, 1951, Wright Papers, Cornell Archives.

61. P. Morrison, personal interview, July 31, 1997. This was even more the case after Dale Corson became chairman at the physics department in 1956. Corson was a member of the physics department and accompanied Morrison to his interrogation by the subcommittee of the Board of Trustees that was to make recommendations on his promotion to a full professorship in 1955.

62. Morrison was born on November 7, 1915, in Somerville, New Jersey. He grew up in a residential suburb of Pittsburgh. He contracted polio as a child, preventing him from entering school until the third grade. He received a B.S. from Carnegie Institute of Technology in Pittsburgh 1936.

63. The recommendation is quoted in the *Memorandum re Professor X.* Malott Papers, Cornell Archives.

64. Interview with S. Schweber, July 31, 1997.

65. Nonetheless, by virtue of his political views, Morrison was under FBI surveillance.

66. Quoted in a letter from Wright to Bache, March 31, 1951. Wright Papers, Cornell Archives.

67. See Schrecker 1986, 151. Schrecker's *No Ivory Tower* is a thorough, well-documented, and reliable account of the impact of McCarthyism on the universities. She has an extensive coverage of the Morrison case on which I have relied and from which I have drawn extensively for the post-1952 period.

68. Truman's proclamation dealt with the need to re-arm to meet the grave external threat to the peace of the world.

69. The newsletter *Counterattack* had been founded by three former FBI agents in 1947 to supplement the government's effort to expose and combat Communist activities.

70. See Bache to Wright, March 27, 1951, which contained a copy of the *Counterattack* article.

71. Arthur H. Dean was a lawyer with Sullivan and Cromwell, a prestigious law firm on Wall Street.

72. Dean to Victor Emanuel, January 6, 1951.

73. T. P. Wright to H. L. Bache, March 31, 1951, Wright Papers, Cornell Archives. The policy of "trying to persuade" Morrison had been worked out in consultations between Day and Wright. Arthur Dean arrived at the same conclusion and supported that policy.

74. In addition to conferring with Bethe, Wright also consulted several other faculty members regarding Morrison.

75. When Morrison testified before the Jenner Senate Subcommittee on Internal Security in April 1953, he was asked whether he had urged the acceptance of the Soviet Union's atomic proposal. He replied: "No. I think I have not done that. I have urged a compromise be sought which would be to the mutual benefit of both powers."

76. Office Files, T. P. Wright, April 17, 1951, Wright Papers, Cornell Archives. Discussions on Philip Morrison with Hans Bethe.

77. Morrison to T. R. Wright, April 18, 1951. Wright Papers, Cornell Archives. After spending a month tracking down all the quotations attributed to him, Morrison, on May 26, 1951, sent Wright a seven-page, single-spaced typed document that consisted of a point-by-point discussion of the material contained on pages 87 through 90 of the April 1 HUAC "Report on the Communist 'Peace' Offensive" in order to prove "that the case they make is in the main a case of innuendo, half-truth, substitution of words for hard facts, given force only by an easy access to the irresponsible sections of the press. If their statements are tested by any of the rules of evidence or scholarship, they fall to the ground."

78. A good part of Morrison's letter is taken up with his explanation of why he will only agree to act in conformity with the spirit of Wright's suggestions. While assenting not to accept sponsorship of any student group which has for its purpose support of the American Peace Crusade, "or indeed any other purpose," Morrison pointed out that there had never been any implication that such a sponsorship must be in the academic field of the faculty sponsor. He "would not want to establish any precedent for such a limitation of right." Concerning "appearing on platforms with avowed or proven Communists," Morrison pointed out that in the more than one hundred public talks that he has given to a wide assortment of audiences with differing political views since coming to Cornell, this has occurred only once or twice. The recent meeting in Washington, over which he had presided and which had been widely reported in the press, at which Paul Robeson had spoken, seems to have been one such event. However, Morrison pointed out, "my most earnest inquiries, direct and indirect, have satisfied me that Mr. Robeson, in spite of the attitude of the press, would not formally meet the definition. I disagree sharply with the fundamental political views expressed by Mr. Robeson, as he would be in an excellent position to confirm, but we share one belief: that some kind of mutual toleration and co-existence is better than an all-out American-Russian war. . . . I am anxious to avoid a repetition of such an affair, and will do my best to see that it is in fact not repeated. I cannot accept, however, the principle of guilt by association, especially when the very 'guilt' of the associate is as tenuous as today's witch hunting will allow." Morrison to Wright, April 18, 1951, Wright Papers, Cornell Archives.

79. Morrison to Wright, ibid.

80. Wright to Morrison, April 23, 1951, ibid. As the further communications between them indicate, they remained in disagreement regarding the American Peace Crusade. Morrison believed that it was "a viable organization, democratically controlled by the consensus of its members and sponsors," and that it formed "a useful means of unifying all sorts of American groups—housewives and union men, academicians, clergymen and farmers—in search for a policy for American peace and security." However, he promised that he would be "especially careful" in any future meetings or public communication of the American Peace Congress to weigh the university's interests, and that he hoped that he

would be able consult Wright from time to time concerning dubious or borderline decisions. Morrison to Wright, April 25, 1951, Wright Papers, Cornell Archives.

81. The committee "appointed to consider the Morrison problem" included S. S. Atwood, John W. MacDonald, R. B. MacLeod, C. C. Murdoch, with Herrell DeGraff as chairman. Murdoch had been appointed to it on the recommendation of Morrison.

82. DeGraff to Malott, January 31, 1953, Malott Papers, Cornell Archives.

83. Atwood et al. to Malott, January 31, 1949, Malott Papers. Two committee members, Professors MacDonald and DeGraff, acting for the committee, tried to interview J. B. Matthews for information concerning the evidence on which his testimony was based. They were unable to do so. They had gone to New York to interview Matthews but were met by John A. Clements, a public-relations director for Hearst Publications and an associate of Matthews in an organization known as Clements Associates, that Clements, Matthews, and former HUAC chairman Martin Dies had formed to gather information about Communists and Communist-front organizations. They "were greeted coldly, in fact with hostility, by Mr. Clements. In most profane language he accused universities and colleges of having been arrogant, shameless, and high handed in harboring communists and fellow travelers." After "having worked off a considerable head of such steam" Clements showed them the "50 three-by-five file cards which carried Morrison's name in the upper left hand corner, a date at the upper right, and cryptic notations below apparently referring to specific locations in the one million files Clements's organization had accumulated." But beyond showing them these cards, Clements refused to produce any evidence about Morrison's activities.

84. These statements were characterized as "unsworn, pertinent" ones in a confidential 1956 "Memorandum re Professor X."

85. Atwood et al. to Malott, Malott Papers, Cornell Archives.

86. Morton Camac, who had known Morrison both at Los Alamos and at Cornell, was at the University of Rochester in 1953. In the spring of that year he decided to leave academia and take a position at the Avco Research Laboratory working with Arthur Kantrowitz. The job required a security clearance. During his interrogation, Morrison's name came up, and the FBI agent who was interviewing him told him that he had just come back from New Hampshire, where he had been trailing Morrison, who was there on vacation, because "the FBI was sure he had gone to the White Mountains to meet Russians there." Interview with Morton Camac, July 30, 1997.

87. Both the *New York Times* and the *Ithaca Sun* reported the hearing in their May 9 edition. The *Times* headline was: ATOMIC BOMB AIDE TELLS OF RED TIES. The article under it stated that Morrison had testified that he was a Communist before he had been "sought out" to work on the atomic bomb project; that atomic security officers and his superiors had known of his Communist background; and that the committee had not asked him and that he did not say when or if he had quit the party. After

the hearing, Morrison told reporters that he was not now and had not been a Communist since he was a young man, and that he was incensed that the subcommittee "did not dare" ask him that question directly, and that this failure had been "deliberate." The *Times* also reported that MIT had issued a statement that Morrison was "engaged as a visitor in fundamental research in physics which has no connection with any Governmental research" and that he had no teaching duty.

88. Schrecker 1986, 157.
89. MacDonald to the Members of the Subcommittee on Academic Problems Arising from Governmental Investigations, August 5, 1953. MacDonald was quoting from the Senate Internal Security Subcommittee (SISS) report. The SISS report had also noted that "in his testimony Morrison stated that certain authorities did know of his record." Malott papers, Cornell Archives.
90. Bethe to MacDonald, November 20, 1953, Bethe Papers, Cornell Archives.
91. See also Schrecker 1986, 157.
92. *Counterattack* 7 (no. 25), June 19, 1953, 3–4.
93. The organization was rumored to be on the attorney general's list. At the meeting, Morrison shared the podium with W.E.B. Du Bois and Pete Seeger.
94. Quoted in Schrecker 1986, 158.
95. Malott to DeGraff, December 8, 1953, Malott Papers, Cornell Archives. Malott added that "Lloyd Smith (the then chair of the physics department) hopes very much we will get something accomplished. It is beginning to bite him on some of his money-raising activities, and he would like to see, I think, some solution."
96. Dale R. Corson to Malott, December 10, 1953, Malott Papers, Cornell Archives.
97. Schrecker 1986, 159.
98. Malott to Morrison, January 28, 1954, Malot Papers, Cornell Archives.
99. Schrecker 1986, 159.
100. The "Memorandum re Professor X" was then forwarded to the executive committee. Throughout the document, Morrison is referred to as X. Its first paragraph contains the following disclaimer: "Since some of the sources of my information are confidential, I am unable to state the sources and in some cases have had to make my own estimate of the reliability of the information. However, I am attempting in this memorandum to be as factual and fair as is possible." Littlewood to members of the ad hoc committee, April 27, 1956, Malott Papers, Cornell Archives.
101. Schrecker 1986, 160.
102. See Schrecker 1994.
103. Report of July 5, 1949, J. R. Oppenheimer, FBI Security File, microform, reel 1 (Wilmington, Del.: Scholarly Resources).
104. For the FBI version of the story, see the report by Charles Brush dated March 15–17, April 4 and 10, 1950. File number 100–3132. J. R. Oppen-

heimer, FBI Security File, microform, reel 1 (Wilmington, Del.: Scholarly Resources).

105. The change in the character of the attacks on Oppenheimer is also reflected in his FBI files. Thus, whereas in 1947 Pitzer told the FBI that he vouches for Oppenheimer's loyalty, in April 1952 he requested an interview with the FBI to state "that he is now doubtful as to the loyalty of Dr. OPPENHEIMER" because of his opposition to the H-bomb. SAC, SF Report 100–17828, April 5, 1952, J. R. Oppenheimer, FBI Security File, microform, reel 1 (Wilmington, Del.: Scholarly Resources).

106. Senator William Jenner, the chairman of the Senate Internal Security Committee, claimed in 1954 that he had discovered voluminous information on Oppenheimer as far back as 1952 and had given it to the proper agencies of the government. For a detailed account of the attacks on Oppenheimer from 1949 until 1953, see Stern 1969 and Major 1971.

107. See, for example, Haberer 1969 and the extensive bibliography therein. Also Shils 1956.

108. I have in mind primarily the members of the school of mathematics and of the school of natural science. Perhaps, within the school of historical studies, George Kennan was the exception, but evidently Oppenheimer did not consult with him. See Kennan 1967.

109. See for example his essay, "An Open House," in Oppenheimer 1954.

CHAPTER 5. NUCLEAR WEAPONS

1. Farrand was president of Cornell when Bethe came there in 1935. Farrand was a member of the Emergency Committee in Aid of Displaced German Scholars that was helping refugee scientists find suitable positions in the United States.

2. Bethe 1964.

3. This is true of all the European scientists who worked at Los Alamos whom I have interviewed: Bethe, Teller, Peierls, Weisskopf, Cyril Smith.

4. None of the following persons told their wives: Robert Bacher, Norris Bradbury, Charles Critchfield, Kenneth Bainbridge, Robert R. Wilson, Robert Marshak.

5. Freeman Dyson made this point in his Preface to the forthcoming Joseph Rotblat Festschrift volume, *The Force of Reason.*

6. The most thorough account of the history of the atomic bomb is Rhodes 1987. See also McGeorge Bundy 1982, 1988. Bundy makes the point that the atomic bomb project in the United States went forward because there existed a small period of time during the summer of 1941 when it could withstand the claims of competing interests. It was during this period that Roosevelt approved the project and essentially wrote a blank check for it. After Pearl Harbor, Roosevelt would not have been able to bestow the lavish resources the project required. Moreover, as Fermi already pointed out in 1945, only the United States was in a position to marshal the requisite technical manpower, funding, and industrial

resources to insure the success of the enterprise, *while waging a total war*. See Fermi 1946.

7. See Gowing 1974, Rhodes 1986, and Peierls 1997. For MAUD, see note 61 of Introduction.

8. Compton was put in charge of the work on the reactor and the bomb design, Lawrence of the project to separate the uranium isotopes by electromagnetic means, and Urey of the gaseous diffusion separation project. See Rhodes 1987.

9. As the bomb explodes, it blows itself up, and this expansion stops the chain reaction before all the fissionable material is used up. The efficiency is the fraction used up in the explosion and determines the energy release. See Serber 1992.

10. See Serber 1998 for his reminiscences of that summer.

11. Joseph Rotblat was one of the very few scientists to leave Los Alamos at that time, and this was in December 1944. A Polish Jew who had emigrated to England before the war, he was one of the sixteen British physicists to join the Los Alamos project in late 1943. At a private dinner in March 1944 at the home of James Chadwick, the head of the British Mission to Los Alamos, Groves had stated that "you realize, of course, that the whole purpose of this project is to subdue the Russians." Groves's statement shocked Rotblat. He had thought that the purpose of Los Alamos had been to cope with the German threat, but evidently its mission had already changed. Groves's declaration initiated Rotblat's doubts about the enterprise. Toward the end of 1944 he was informed by Chadwick, who was close to British intelligence, that Germany had stopped work on the bomb. He accordingly decided to end his participation in the project. Rotblat had believed that "if the U.S. and Great Britain had developed the bomb, then even if Germany had it, we could have made the Germans give up using it. It was the idea of deterrence" (Rotblat 1995). Groves testified at the Oppenheimer hearings in 1954 that the views regarding the aim of the Los Alamos project expressed at the Chadwick dinner had shaped his policy regarding security matters at Los Alamos. "There was never from about 2 weeks from the time I took charge [of the atomic bomb] project any illusion on my part but that Russia was our enemy and the project was conducted on that basis." In Oppenheimer 1970, 173.

12. Philip Morrison, who served as one of the scientists assigned to Groves to evaluate the data that Goudsmit had seized for the Alsos mission as the Allies swept through France, had came to the conclusion in December 1944 that Germany was not working on a bomb. For the Alsos mission, see Goudsmit 1947.

13. Teller 1962, 13. The Frank Report issued in Chicago in June 1945 had suggested that perhaps with a demonstration the Japanese would see that the best thing to do would be to have the war end without the bomb being used against them.

14. Quoted in Sherwin 1975, 145.

15. Lawrence to Darrow, August 17, 1945. Darrow Papers, American Institute of Physics (AIP), College Park, Md.

> I made the proposal [of a demonstration of the bomb] briefly in the morning session of the Secretary of War's committee, and during luncheon Justice Byrnes (now Secretary of State) asked me further about it, and it was discussed at some length. . . .

> I am sure it was given serious consideration by the Secretary of War and his Committee, and gather from the discussion that the proposal to put on a demonstration did not appear to be desirable because, as you say, (a) the number of people killed by the bomb would not be greater in order of magnitude than the number already killed in fire raids, and (b) Oppenheimer could think of no demonstration that would be sufficiently spectacular to convince the Japs that further resistance was useless. Oppenheimer felt, and the feeling was shared by Groves and others, that the only way to put on a demonstration would be to attack a real target of built-up structures.

> In view of the fact that two bombs ended the war, I am inclined to feel that they made the right decision. Surely many more lives were saved by shortening the war than were sacrificed as a result of the bombs. Further, it goes without saying that all of us hope and pray that there will never be an occasion to use another one. The world must realize that there can never be another war.

> As regards criticisms of science and scientists, I think that is a cross we will have to bear, and I think in the long run the good sense of everyone the world over will realize that in this instance, as in all scientific pursuits, the world is better as a result.

16. The report of the Science Panel, "Recommendations on the Immediate Use of Atomic Weapons," June 16, 1945, can be found in Sherwin 1987, 304–305. It is interesting to note that the scientific panel added the following cautionary statement to their recommendation: "We, as scientific men, have no claim to special competence in solving the political, social, and military problems which are presented by the advent of atomic power."

17. Report by D. E. Todd dated April 5, 1947. J. R. Oppenheimer, FBI Security File, microform, reel 1 (Wilmington, Del.: Scholarly Resources).

18. Bethe, during the spring of 1945, had been responsible for assessing the "Expected damage of the Gadget." Damage due to the fact that the temperature generated in an atomic explosion is in the million degree range; due to the extended duration of the shock waves generated; due to the intense neutron radiation emitted from the explosion that diffuses over a range of a few kilometers; due to the radioactive fission products released.

19. The Jornada del Muerto is a ninety-mile stretch of bleak desert in central New Mexico. Its name is variously translated from the Spanish as either "Journey of Death" or "Trail of the Dead Man."

20. T. R. Farrell, report, appendix in Groves 1963.

21. Quoted in Szasz 1984, 89.

22. Also in the observation post with Bethe was the *New York Times* reporter William L. Laurence, whom Bethe knew. After the bomb went off, Bethe recalls a tremendous spectacle, and after about a minute, a tremendous roar. A terribly anguished Laurence asked Bethe: "For heaven's sake, what was that?" Bethe explained to him the difference between the velocity of sound and the velocity of light. Laurence was relieved. After a while came the reverberations from the mountains, some ten miles behind the bunker. According to Bethe, "The acoustic spectacle was not quite as massive as the visual one, but certainly quite impressive."

    Bethe had been in charge of estimating the size of the bomb and had arrived at a figure of eight kilotons. After the blast he had the responsibility of evaluating the magnitude of the explosion from photographs and other data that had been collected. He did this for three days at the headquarters near ground zero and came up with the figure of twenty kilotons as the TNT equivalent of the Trinity bomb.

23. See Serber 1998 for the letters he wrote to his wife from Nagasaki and Hiroshima. It should be added that in *Peace and War* Serber exhibits both insightfulness into and a disturbing detachment from the world around him. All of Oppenheimer's students in the 1930s became politically engaged: they were concerned about fascism, Loyalist Spain, the plight of migrant workers. Serber remained aloof. Both he and Philip Morrison went to Tinian, and to Hiroshima and Nagasaki. As we have seen, Morrison was deeply affected by what he had beheld and subsequently devoted considerable efforts—at great risk to himself during the McCarthy era—to find all means possible to avert a nuclear war. Serber came back from the assignment and also gave seminars at Los Alamos about what he had seen. In one of them he showed a photograph he had taken on the outskirts of Nagasaki of a horse which had all its hair burnt off on one side, but whose other side was perfectly normal. In his talk Serber remarked "that the horse was happily grazing, and Oppie scolded [him] for giving the impression that the bomb was a benevolent weapon" (Serber 1998, 129).

24. Oppenheimer to Stimson, August 17, 1945. Oppenheimer Papers, Archives Division, Library of Congress.

25. For Oppenheimer's role in drafting atomic energy legislation, see Bernstein 1974. Bethe has been deeply absorbed with the problems raised by nuclear weapons ever since August 1945, and has frequently written about the perils they pose. Some of these writings have been collected in Bethe 1991.

26. See York 1989; Rhodes 1995.

27. For the oscillations of Oppenheimer regarding the building of the H-bomb, see Galison and Bernstein 1989, which is the best overview on

the positions of American scientists and the decision to build the Super-bomb. See also Rhodes 1995.

28. See Hershberg 1988 and 1993. One of the valuable aspects of Hershberg's book *James B. Conant* is the light it sheds on the Conant-Oppenheimer relationship. The rapport and mutual affinity between Conant and Oppenheimer furnishes a clue as to why in 1933 Conant gave up chemistry to assume the presidency of Harvard, and Oppenheimer gave up physics to become the director of the Institute for Advanced Study. Until 1940 Oppenheimer was probably the most brilliant theoretical physicist in the United States. Although he was intellectually as gifted as Heisenberg, Schrödinger, Pauli, and Dirac—the creators of the new quantum mechanics—he felt himself not to be in their class in terms of creativity. Similarly, Conant probably was the best physical organic and natural product chemist in the United States during the twenties. He had recognized that the next important set of developments in organic chemistry would lie in the biological realm, and he had been working on the problem of the structure of chlorophyll. But, like Oppenheimer, he probably had felt that he was not as creative as the very best chemists in Europe. Thus, both of them gravitated toward administration—Oppenheimer assuming the directorship of the Institute after Los Alamos—and toward statesmanship.

29. Galison and Bernstein 1989, 291, n. 84.

30. Oppenheimer was consistent in this position. He took an active part in Project Vista, which was established at Cal Tech during 1950–51 "to study some of the problems of ground and air tactical warfare, especially as they relate to the defense of Western Europe." The letter to the secretaries of the army, navy, and air force that transmitted the report—the letter was dated February 4, 1952, and was probably drafted by Oppenheimer—asserted that "we believe that the United States, in collaboration with its allies in the North Atlantic Treaty Organization, can prevent the military conquest of Western Europe by the Soviet Union—and can do this in 1952 if necessary—if we try." The report recommended "the tactical employment of our atomic weapon resources," which the participants in the project believed "holds outstanding promise." They recommended specific weapons design and use and stressed that "if the allied nations have the strength to resist successfully an attack by the Soviet Union, this attack will probably not take place and war can be prevented." In the fall of 1951, Oppenheimer traveled to Paris to convince General Eisenhower, then the commander in chief of NATO, that the use of tactical weapons would form an effective defense against a Soviet attack on Western Europe. The recommendations of Project Vista became the basis of NATO strategy. A copy of the letter of transmittal and a heavily censored copy of the report can be found in the Bacher Papers, Cal Tech Archives, Pasadena, Calif. See also chapter 12 of Dyson 1984.

31. Oppenheimer 1947, 8. Bethe shared these views. In fact, he had never been captivated by the communist vision, not even during the 1930s.

32. Oppenheimer had been deeply disturbed by Curtis LeMay, the head of the Air Force Strategic Command, and his plans for the use of nuclear weapons.
33. Quoted in Galison and Bernstein 1989, 307, and n. 144 therein.
34. Quoted in Davis 1968, 330.
35. Bethe assumed only an "official" capacity in the fall of 1956, when he was appointed to PSAC.
36. Bethe in Oppenheimer 1970, 328.
37. See the petition for a pledge by the United States not to be the first to use an H-bomb signed by Bethe and eleven other prominent physicists at the New York meeting of the American Physical Society at the end of January 1950. The list of signatories included S. K. Allison, director of the Institute for Nuclear Studies, University of Chicago; K. T. Bainbridge, Harvard University; H. A. Bethe, Cornell University; C. C. Lauritsen, Cal Tech; G. B. Pegram, Columbia University; F. Seitz, University of Illinois; M. A. Tuve, Carnegie Foundation; and V. F. Weisskopf, MIT. "Let Us Pledge Not to Use an H-bomb First," *Bulletin of Atomic Scientists* 6/3 (1950): 75.
38. Quoted in Galison and Bernstein 1989, 309.
39. Bethe to Weisskopf, February 14, 1950, Box 9, Fan Mail 1938–50, Bethe Papers, Cornell Archives.
40. The article in *Scientific American* was due to appear earlier, but the AEC confiscated the printing plates fearing that the article contained sensitive information. The version that was published in April is a very slightly amended version of the original manuscript.
41. Bethe 1950a.
42. For a description of the Super, see Truslow and Smith 1983. For the mechanism suggested by Ulam, Teller, and de Hoffman, see Hewlett and Duncan 1972; Ulam 1976; and especially Rhodes 1995 and Hansen 1988, 1995.
43. See Bethe 1950a.
44. Bethe to Bradbury, February 14, 1950, Los Alamos Archives.
45. In 1952 Teller had claimed that Oppenheimer had used his position to influence people not to work on the H-bomb from 1949 until 1951, and he pointed to Bethe as one person so influenced. In May 1952, the FBI interviewed Bethe. During the interview Bethe denied the allegation: "He was never approached by any individual in an effort to influence him not to work on the H-bomb." Bethe also stated that "he still believes our only justification for building the H-bomb is to deter another nation from using it against us." FBI, Oppenheimer Security File 100–17828, May 19, 1952.
46. See also "The Hydrogen Bomb," *Foreign Policy Reports* 26, no. 8, September 1, 1950. This is a condensed version of Bethe 1950a,b.
47. I thank Priscilla McMillan for providing me with a transcript of Bethe's testimony.
48. "When I started working on the thermonuclear in Summer 1950, I was hoping to prove that thermonuclear weapons could not be made."

Bethe 1954, declassified in the early 1980s and published as Bethe 1982a, 51; also in York 1989.

49. Bethe to Gordon Dean (the chair of the AEC), May 23, 1952, CD 471.6, Office of the Secretary of Defense Records, RG 330, National Archives, Washington, D.C. Ulam had gone to Teller with the idea that the conventional A-bomb be used to compress materials to high densities, rather than to generate extremely high temperatures. Teller, recognizing the importance of high densities, then suggested that the X rays emitted in A-bomb explosions be used to compress deuterium and tritium rather than heating them, i.e., that A-bomb-generated radiation implosion might be capable of achieving the required extreme densities over large volumes. See Teller 1955.

50. Bethe 1982a, 51; emphasis in original.

51. Edson 1968, 125.

52. Ibid.

53. Jungk 1958, 291.

54. Bethe to Gordon Dean, September 9, 1952, Department of Energy Archive, Gordon Dean Papers, Washington, D.C.

55. Ibid. I am indebted to Priscilla McMillan for making her copy of this letter available to me.

56. Anders 1987, 226–231.

57. See Bernstein 1989.

58. The panel had made five general recommendations: (1) that nuclear secrecy be reduced and that there be more candor with the American people; (2) that more communication be established with the Allies about nuclear weapons, strategy, and defense; (3) that a continental air defense system against a Soviet knockout nuclear attack be investigated; (4) that the United States withdraw from United Nations nuclear disarmament discussions because of their rigidity and insincerity; (5) that improved communications with the Soviet Union be established. Bernstein 1989, 154.

59. Bernstein 1989.

60. I stress "*at that time*" since new simulation methods and high-speed computing have made actual testing marginal.

61. Bethe to Gordon Dean, May 23, 1952, CD 471.6, Records of the Secretary of Defense, RG 330, National Archives. According to David Holloway, the leading Western scholar of the Soviet nuclear weapons program, the October 1952 tests did "speed up Soviet efforts," and thereafter the tempo of work increased. Holloway 1980, 195.

62. Bethe 1982a, 53.

63. Sakharov's decision to participate in the Soviet hydrogen bomb project was motivated by his conviction that the world would be safer with a socialist bomb to balance a capitalist one: "I had no doubts as to the vital importance of creating a Soviet superweapon—for our country and for the balance of power throughout the world." But by the 1960s, after seeing his work initiate an arms race that put tens of thousands of nuclear bombs in the arsenals of the United States and the Soviet Union,

Sakharov became ever more concerned about the dangers of a thermonuclear war. Deeply troubled by the harmful effects of radioactive fallout from continued testing of nuclear devices in the atmosphere, and disillusioned by the rejection of his recommendation to the Soviet leadership in 1961 not to resume atmospheric testing after a three-year moratorium, Sakharov became an outspoken dissident of the Soviet regime and a vigorous opponent of the nuclear arms buildup that had reached mindless proportions.

For an account of the Soviet work on their hydrogen bomb, see Rhodes 1995 and the review of this book by Genadi Gorelik in *Physics Today.*

64. Bethe 1982, 53.
65. For a dissenting view concerning Bethe's assertion, see Dyson 1984, 36–44. With the increase in accuracy of missiles, MIRV (Multiple Independently-targeted Reentry Vehicles) and cruise missiles made it possible to deliver low-yield weapons in large numbers accurately and cheaply—and, according to Dyson, made high-yield hydrogen bombs obsolete. Therefore, "the hydrogen bomb has become almost irrelevant," meaning that "a hypothetical world without hydrogen bombs [i.e., had the GAC recommendations been adopted] would not have been appreciably different from those which we are now deploying" (Dyson 1984, 38–39).
66. Bethe 1964, 3.
67. H. A. Bethe, *FAS Public Interest Report,* September–October 1995. In issuing the statement, Bethe was also responding to Teller's call for a third generation of nuclear weapons.
68. For a more radical proposal, see MacKenzie and Spinardi 1996.
69. See Bernstein 1982, Kunetka 1982, Stern 1969, and York 1989.
70. Rieff 1969, 333.
71. Killian 1977, 111.
72. Ibid., 111, 116.
73. During its first few years, nearly two hundred "outside" scientists and engineers served on PSAC panels. For a list of the panels (space, missiles, etc.) and their reports, see Killian 1977, 119–150, and Maier on p. lvi of Kistiakowsky 1976.
74. The irradiation of the Japanese fishermen aboard the *Lucky Dragon,* which was a few hundred miles downwind from the test area of the American "Bravo" shot on March 1, 1954, precipitated a global opposition to nuclear testing. The bomb used in this particular test was a "dirty" (fission-fusion-U^{238} fission bomb).
75. See Hewlett and Holl 1989 for a detailed account of Bethe's involvement from 1957 until 1963 that made possible the ratification of the test ban treaty.
76. Hewlett and Holl 1989, 472.
77. Killian 1977, 154–155.
78. See Divine 1978 and Killian 1977 for details of the conference. Bethe had made calculations that indicated that the "big hole" method of muffling tests would not work. His later calculations convinced him that the orig-

inal ones were in error, and "with great courage he acknowledged this openly to the Soviets at the technical session of the Diplomatic Conference in Geneva." Killian 1977, 155.

79. For an account of these events, see Divine 1978, Kistiakowsky 1976, and Killian 1977.

80. Bethe 1960.

CHAPTER 6. ON SCIENCE AND SOCIETY

1. The lectureship had been established in 1944, but the dislocation of normal activities due to the war had made it impossible to inaugurate the series until 1946.
2. Oppenheimer 1947, 3.
3. The words were those of Secretary of War Henry Stimson when awarding him the Medal of Merit.
4. Oppenheimer 1947, 6.
5. The other things that Oppenheimer was alluding to which may come out of science and ease man's labor are "the things which . . . will shorten his working day and take away the most burdensome part of his effort, which will enable him to communicate, to travel, and to have a wider choice both in the general question of how he is to spend his life and in the specific question of how he is to spend his time of leisure" (Oppenheimer 1947, 10).
6. The points regarding skepticism and the other attributes of the scientific community are reiterated and amplified in a lecture entitled "The Encouragement of Science," which he delivered in 1950. Oppenheimer 1955, 114–116.
7. See also ibid., 115.
8. Oppenheimer 1958, 3.
9. Ibid., 3–4.
10. Rouze 1962, 147.
11. Oppenheimer had done so before C. P. Snow made the notion of the two cultures popular; see Snow 1959. See, in particular, Oppenheimer's 1953–54 lectures on "The Scientist in Society" and "Prospects in the Arts and Sciences" in Oppenheimer 1955, 119–146.
12. Oppenheimer 1960, 25.
13. Barton Papers, Box 119, Folder 3, Niels Bohr Library, AIP, College Park, Md.

EPILOGUE

1. Eliot 1952, 68–69.
2. See Beer 1975, 35ff. and the references therein.
3. This was the concluding lecture in a series dedicated to "Man's Right to Knowledge and the Free Use Thereof" on the occasion of Columbia University's bicentennial.

4. Recall the reordering of the Enlightenment of the former unified religion and metaphysics into three autonomous spheres.
5. Oppenheimer 1955, 144–145.
6. Recall the words used by Oppenheimer in thanking President Lyndon Johnson upon being awarded the Enrico Fermi Award of the Atomic Energy Commission in 1963: "I think it just possible . . . that it has taken some charity and some courage for you to make this award today."
7. Kennan 1967, 56.

Bibliography

Acheson, D. 1969. *Present at the Creation*. New York: Norton.

Adler, F. 1881. *The Anti-Jewish Agitation in Germany: Larger Tolerance*. Two addresses delivered at Chickering Hall, December 19 and 26, 1880. New York: Lehmaier.

Adler, F. 1892. *The Moral Instruction of Children*. Vol. 21 of the International Education Series, ed. William T. Harris. New York: D. Appleton.

Adler, F. 1902. "A Critique of Kant's Ethics." *Mind* 11:180–245.

Adler, F. 1904. "The Problem of Teleology." *International Journal of Ethics* 14:265–280.

Adler, F. 1918. *An Ethical Philosophy of Life*. New York: D. Appleton.

Agar, J., and B. Balmer. 1998. "British Scientists and the Cold War: The Defence Research Policy Committee and Information Networks, 1947–1963." *Historical Studies in the Physical and Biological Sciences* 28/2:209–252.

Allen, J. 1970. *March 4: Scientists, Students and Society*. Cambridge, Mass.: MIT Press.

Altmann, A. 1973. *Moses Mendelsohn: A Biographical Study*. University, Ala.: The University of Alabama Press.

Anders, R. M., ed. 1987. *Forging the Atomic Shield*. Chapel Hill: University of North Carolina Press.

Bacher, R. F. 1972. "Robert Oppenheimer (1904–1967)." *Proceedings of the American Philosophical Society* 116/4:279–293.

Badash, L., J. O. Hirschfelder, and H. P. Broida, eds. 1980. *Reminiscences of Los Alamos, 1943–1945*. Dordrecht, Holland: Reidel.

Bainbridge, K. T. 1975. "A Foul and Awesome Display." *Bulletin of the Atomic Scientists* 32/5:46.

Bainbridge, K. T. 1976. "Trinity." Los Alamos, N.M.: Los Alamos National Laboratory.

Baudelaire, C. 1964. *The Painter of Modern Life, and Other Essays*. Translated and edited by Jonathan Mayne. London: Phaidon.

Bauman, Z. 1997. *Postmodernity and Its Discontents*. New York: New York University Press.

Beer, S. 1975. *Platform for Change*. New York: John Wiley and Sons.

Bender, T. 1993. *Intellect and Public Life: Essays on the Social History of Academic Intellectuals in the United States*. Baltimore: Johns Hopkins University Press.

Bennett, J. 1960. *Four Metaphysical Poets: Donne, Herbert, Vaughn, Crashow*. New York: Vintage Books.

Berlin, I. 1980. "Benjamin Disraeli, Karl Marx and the Search for Identity." In *Against the Current: Essays in the History of Ideas*. New York: Viking Press.

Berlin, I. 1996. *The Sense of Reality*. New York: Farrar, Straus and Giroux.

Bernstein, B. J. 1974. "The Quest for Security: American Foreign Policy and International Control of Atomic Energy, 1942–1946." *Journal of American History* 60/3:1003–1044.

Bernstein, B. J. 1982. "In the Matter of J. Robert Oppenheimer." *Historical Studies in the Physical Sciences* 12/2:192–252.

Bernstein, B. J. 1983. "The H-Bomb Decisions: Were They Inevitable?" In B. Brodie, M. D. Intriligator, and R. Kolkowicz, eds., *National Security and International Stability*, pp. 327–356. Cambridge, Mass.: Oelgeschlager, Gunn and Hain.

Bernstein, B. J. 1988. "Four Physicists and the Bomb: The Early Years, 1945–1950." *Historical Studies in the Physical Sciences* 18/2:231–262.

Bernstein, B. J. 1989. "Crossing the Rubicon: A Missed Opportunity to Stop the H-Bomb?" *International Security* 14/2:132–160.

Bernstein, J. 1975. "Rabi." *New Yorker*, October 13 and October 20, 1975.

Bernstein, J. 1979. *Hans Bethe, Prophet of Energy.* New York: Basic Books.

Bethe, H. A. 1933. "Eins und zwei-Elektron Probleme." In *Handbuch der Physik*, vol. 24/2. Berlin: Geiger/Scheel.

Bethe, H. A. 1950a. "The Hydrogen Bomb." *Bulletin of the Atomic Scientists* 6/4:99–104.

Bethe, H. A. 1950b. "The Hydrogen Bomb: II." *Scientific American* 182/4:21–25.

Bethe, H. A. 1954. "Comments on the History of the H-Bomb." Los Alamos Archives. Declassified and published as Bethe 1982a.

Bethe, H. A. 1958. Review of Jungk's *Brighter than a Thousand Suns. Bulletin of the Atomic Scientists* 12:426–429.

Bethe, H. A. 1960. "The Case for Ending Nuclear Tests." *Atlantic Monthly* 206 (August): 43–51.

Bethe, H. A. 1962. "Science." An Interview by D. McDonald with Hans Bethe. Santa Barbara: Center for the Study of Democratic Institutions.

Bethe, H. A. 1963. "The Social Responsibilities of Scientists and Engineers." *The Cornell Engineer*, October.

Bethe , H. A. 1964. "The Social Responsibilities of Scientists and Engineers." *SSRS Newsletter* 39 (February and March):1964.

Bethe, H. A. 1968a. "Oppenheimer." *Biographical Memoirs of Fellows of the Royal Society* 14:390–416.

Bethe, H. A. 1968b. "Energy Production in Stars." *Physics Today* 20 (September):36–44.

Bethe, H. A. 1982a. "Comments on the History of the H-Bomb." *Los Alamos Science* 3/3:42–54.

Bethe, H. A. 1982b. "The Bradbury Years." *Los Alamos Science* 4/1:26–67.

Bethe, H. A. 1986. *Basic Bethe.* Seminal articles on nuclear physics from the *Reviews of Modern Physics, 1936–37* by H. A. Bethe, R. F. Bacher, and M. S. Livingston. Los Angeles: Tomash Publishers.

Bethe, H. A. 1991. *The Road from Los Alamos.* New York: Touchstone Books.

Bethe, H. A., and J. R. Oppenheimer. 1946. "Reaction of Radiation on Electron Scattering and Heitler's Theory of Radiation Damping." *Physical Review* 70:451–458.

Birmingham, S. 1967. *"Our Crowd": The Great Jewish Families of New York.* New York: Harper and Row.

Blumberg, S. A., and G. Owens. 1976. *Energy and Conflict: The Life and Times of Edward Teller.* New York: Putnam's Sons.

Bohr, N. 1972–. *Collected Works.* Amsterdam: North-Holland.

Bohr, N., and L. Rosenfeld. 1933. "Zur Frage des Messbarkeit der electromagnetischen Feldgrössen." *Kgl. Danske vidensk Selskab Mat. Fys. Medd.* 12, no. 8.

Born, M., and J. R. Oppenheimer. 1927. "Zur Quantentheorie der Molekülen." *Annalen der Physik*, 4th series, 84:457–484.

Boyer, P. 1985. *By the Bomb's Early Light: American Thought and Culture at the Dawn of the Atomic Age.* New York: Pantheon Books.

Breit, G. 1932. "Quantum Theory of Dispersion." *Reviews of Modern Physics* 4:504–545.

Bridgman, P. W. 1947. "Scientists and Social Responsibility." *Scientific Monthly* 65, no. 2:148–154.

Brown, L. M., and L. Hoddeson, 1983. *The Birth of Particle Physics.* Cambridge, U.K.: Cambridge University Press.

Brown, L. M., M. Dresden, and L. Hoddeson, eds. 1989. *Pions to Quarks. Particle Physics in the 1950s.* Cambridge, U.K.: Cambridge University Press.

Bruford, W. H. 1975. *The German Tradition of Self-Cultivation.* Cambridge, U.K.: Cambridge University Press.

Bundy, McG. 1982. "The Missed Chance to Stop the H-Bomb." *New York Review of Books*, May 13, 1982, 13–21.

Bundy, McG. 1988. *Danger and Survival.* New York: Random House.

Burrow, J. W. 1966. *Evolution and Society: A Study in Victorian Social Theory.* London: Cambridge University Press.

Bush, V. 1960. *Science: The Endless Frontier.* Reissue of 1945 report as part of the 10th anniversary observance of the National Science Foundation. Washington, D.C.: National Science Foundation.

Camus, A. 1962. *The Rebel: An Essay on Man in Revolt.* With a foreword by Herbert Read. A revised and complete translation of *L'Homme revolté* by Anthony Bower. New York: Random House.

Carlson, J. I., and J. R. Oppenheimer. 1937. "On Multiplication Showers." *Physical Review* 81:220–231.

Carr, R. K. 1952. *The House Committee on Un-American Activities, 1945–1950.* Ithaca, N.Y.: Cornell University Press.

Chandrasekhar, S., G. Gamow, and M. A. Tuve. 1938. "The Problem of Stellar Energy." *Nature* 141:982.

Charle, C., J. Schriever, and P. Wagner. 1977. *Transnational Intellectual Networks and the Cultural Logics of Nations.* Oxford: Berghahn Books.

Chevalier, H. 1959. *The Man Who Would Be God.* New York: Putnam.

Chevalier, H. 1965. *Oppenheimer: The Story of a Friendship.* New York: G. Braziller.

Clarke, E. 1997. *Theory and Theology in George Herbert's Poetry.* Oxford: Clarendon Press.

Cohen, N. W. 1978. "American Jewish Reactions to Anti-Semitism in Western Europe, 1875–1900." *Proceedings of the American Academy of Jewish Research* 45:29–65.

Condon, E. U., and P. M. Morse. 1929. *Quantum Mechanics.* New York: McGraw-Hill.

Conser, W. H., and S. B. Twiss, eds. 1998. *Religious Diversity and American Religious History: Case Studies in Traditions and Cultures.* Athens: University of Georgia Press.

Dallet, J. 1938. *Letters from Spain to His Wife.* New York: Workers Library Publication.

Daston, L. 1995. "The Moral Economy of Science." *Osiris* 10:1–7.

Davies, P., ed. 1989. *The New Physics.* Cambridge, U.K.: Cambridge University Press.

Davis, N. P. 1968. *Lawrence and Oppenheimer.* New York: Simon and Schuster.

Dayton, B., D. Lal, N. Lund, P. Morrison, and H. W. Schnopper. 1993. "Bernard Peters." *Physics Today* 46 (December):64–65.

Debye, P., ed. 1928. *Probleme der modernen Physik.* Leipzig: Verlag S. Hirzel.

Dewey, J. 1899. *The School and Society.* Chicago: University of Chicago Press.

Dewey, J. 1900. *The School and Society.* Being three lectures by John Dewey . . . supplemented by a statement of the University elementary school. 3d ed. Chicago: University of Chicago Press.

Dewey, J. 1937a. *The Case of Leon Trotsky.* Report of hearings on the charges made against him in the Moscow trials, by the preliminary commission of inquiry, John Dewey, chairman, Carleton Beals (resigned), Otto Ruehle, Benjamin Stolberg, Suzanne La Follette, secretary. New York and London: Harper and Brothers.

Dewey, J. 1937b. *Truth Is on the March.* Report and remarks on the Trotsky hearings in Mexico. New York: American Committee for the Defense of Leon Trotsky.

Dewey, J. 1939. "Creative Democracy—The Task before Us." In R. Rattner, ed., *The Philosophy of the Common Man. Essays in Honor of John Dewey to Celebrate His Eightieth Birthday,* pp. 220–228. New York: Putnam.

Dewey, J. 1966. *Lectures in the Philosophy of Education, 1899.* Edited and with an introduction by R. G. Archambault. New York: Random House.

Dinnerstein, L. 1994. *Antisemitism in America.* New York: Oxford University Press.

Dirac, P.A.M. 1928. "The Quantum Theory of the Electron." *Proceedings of Royal Society London* A 117:610–624.

Dirac, P.A.M. 1930. "A Theory of Electrons and Protons." *Proceedings of Royal Society London* A 119:360–365.

Dirac, P.A.M. 1931. "Quantized Singularities in the Electromagnetic Field." *Proceedings of Royal Society London* A 133:60–72.

Dirac, P.A.M. 1977. "Recollections of an Exciting Era." In *History of Twentieth Century Physics,* edited by C. Weiner. New York: Academic Press.

Dirac, P.A.M. 1978. "The Prediction of Antimatter." The First H. R. Crane Lecture, University of Michigan, Ann Arbor.

Divine, R. A. 1978. *Blowing on the Wind: The Nuclear Test Ban Debate, 1954–1960.* New York: Oxford University Press.

Douglas, M. 1966. *Purity and Danger: An Analysis of Concepts of Pollution and Taboo.* New York: Praeger.

Dyson, F. J. 1979. *Disturbing the Universe.* New York: Harper and Row.

Dyson, F. J. 1983. "Bombs and Poetry." In *Values at War: Selected Tanner Lectures on the Nuclear Crisis* by Freeman Dyson, Raymond Aron, and Joan Robinson, edited by Sterling M. McMurrin. Salt Lake City: University of Utah Press.

Dyson, F. J. 1984. *Weapons and Hope.* New York: Harper and Row.

Dyson, F. J. 1989. Preface in Oppenheimer 1989.

Dyson, F. J. 1992. *From Eros to Gaia.* New York: Pantheon.

Dyson, F. 1997. "Technology and Social Justice." The Fourth Louis Nizer Lecture on Public Policy, November 5, 1997, Carnegie Council on Ethics and International Affairs.

Edson, L. 1968. "Scientific Man for All Seasons." *New York Times Magazine,* March 10, 1968, 29, 122–127.

Eliot, T. S. 1952. *The Complete Poems and Plays.* New York: Harcourt, Brace and Company.

Elsasser, W. 1978. *Memoirs of a Physicist in the Atomic Age.* New York and Bristol: Science History Publications.

Erwin, E., S. Gendin, and L. Kleiman. 1994. *Ethical Issues in Scientific Research: An Anthology.* New York: Garland.

Fermi, E. 1932. "Quantum Theory of Radiation." *Reviews of Modern Physics* 4:887–932.

Fermi, E. 1946. "The Development of the First Reacting Pile." *Proceedings of the American Philosophical Society* 90:20–24.

Feynman, R. P. 1955. "The Value of Science." *Engineering and Science* 19 (December):13–22.

Feynman, R. P. 1958. "The Value of Science." In Hutchings 1958, 260–267.

Feynman, R. P. 1993. *The Meaning of It All: Thoughts of a Citizen Scientist.* Reading, Mass.: Addison-Wesley.

Fitch, V. L., D. R. Marlow, and M.A.E. Dementi, eds. 1997. *Critical Problems in Physics: Proceedings of a Conference Celebrating the 250th anniversary of Princeton University.* Princeton, N.J.: Princeton University Press.

Fleming, D., and B. Baylin, eds. 1969. *The Intellectual Migration.* Cambridge, Mass.: Harvard University Press.

Forman, P. 1987. "Behind Quantum Electronics: National Security as Basis for Physical Research in the United States, 1940–1960." *Historical Studies in Physical and Biological Sciences* 18:152–268.

Forman, P. 1989. "Social Niche and Self-image of the American Physicists." Unpublished manuscript.

Forman, P. 1998. "Molecular Beam Measurements of Nuclear Moments before Magnetic Resonance. Part 1. I. I. Rabi and Deflecting Magnets to 1938." *Annals of Science* 55/2:111–160.

Foucault, M. 1984. "What Is Enlightenment?" In Rabinow 1984. Translated by Catherine Porter.

Friedrich, C. J., ed. 1947. *The Philosophy of Kant.* New York: Modern Library.

Friess, H. L. 1981. *Felix Adler and Ethical Culture: Memories and Studies.* New York: Columbia University Press.

Frisch, O. R. 1967. "How It All Began." *Physics Today* 20/10:42–52.

Frisch, O. R. 1979. *What Little I Remember.* Cambridge, U.K.: Cambridge University Press.

Furry, W., and J. R. Oppenheimer. 1934. "On the Theory of the Electron and Positron." *Physical Review* 45:245–262.

Galison, P. 1987. *How Experiments End.* Chicago: University of Chicago Press.

Galison, P., and B. J. Bernstein. 1989. "In Any Light: Scientists and the Decision to Build the Superbomb, 1942–1954." *Historical Studies in Physical and Biological Sciences* 19/2:267–347.

Galison, P., and B. Hevly. 1992. *Big Science: The Growth of Large-Scale Research.* Stanford, Cal.: Stanford University Press.

Gamow, G. 1928. "Zur Quantentheorie des Atomkernes." *Zeitschrift f. Physik* 51:204–212.

Georgi, H. 1989. "Effective Quantum Field Theories." In Davies 1989, 446–457.

Gerber, D. A., ed. 1986. *Anti-Semitism in American History.* Urbana: University of Illinois Press.

Gilman, S. 1986. *Jewish Self-Hate.* Ithaca, N.Y.: Cornell University Press.

Gilpin, R. 1962. *American Scientists and Nuclear Weapons.* Princeton, N.J.: Princeton University Press.

Gleick, J. 1992. *Genius: The Life and Times of Richard Feynman.* New York: Pantheon.

Goldberg, S. 1992. "Groves Takes the Reins." *Bulletin of Atomic Scientists* 12:32–39.

Goldberg, S. 1995. "Groves and Oppenheimer: The Story of a Partnership." *Antioch Review* 53(4):482–493.

Goodchild, P. 1981. *J. Robert Oppenheimer: Shatterer of Worlds.* Boston: Houghton Mifflin.

Goodlander, M. R. 1922. *Education through Experience: A Four Year Experiment in the Ethical Culture School.* 2nd ed. New York: Published by the Bureau of Educational Experiments in Co-operation with the Parents and Teachers Association, Ethical Culture School.

Goudsmit, S. A. 1947. *Alsos.* New York: H. Schuman.

Gowing, M. 1974. *Independence and Deterrence: Britain and Atomic Energy.* London: Macmillan.

Graham, L. 1981. *Between Science and Values.* New York: Columbia University Press.

Groves, L. R. 1962. *Now It Can Be Told: The Story of the Manhattan Project.* New York: Harper.

Gurney, R. N., and E. U. Condon. 1929. "Quantum Mechanics and Radioactive Disintegration." *Physical Review* 33:127–140.

Haberer, J. 1969. *Politics and the Community of Science.* New York: Van Nostrand Reinhold.

Habermas, J. 1986. "Taking Aim at the Heart of the Present." In Hoyt 1986, 103–108.

Hahn, O. 1958. "The Discovery of Fission." *Scientific American* 198/2:76–84.

Hales, P. B. 1997. *Atomic Spaces: Living on the Manhattan Project.* Urbana and Chicago: University of Illinois Press.

Handlin, O. 1959. *John Dewey's Challenge to Education.* New York: Harper.

Hansen, C. 1988. *U.S. Nuclear Weapons: The Secret History.* Arlington, Tex.: Aerofax.

Hansen, C., ed. 1995. *The Swords of Armageddon: U.S. Nuclear Weapons Development since 1945.* Sunnyvale, Cal.: Chukelea Publications (one computer optical disk and one booklet).

Harwood, J. 1987. "National Styles in Science: Genetics in Germany and in the United States between the World Wars." *Isis* 78:390–414.

Harwood, J. 1993. *Styles of Scientific Thought: The German Genetics Community, 1900–1933.* Chicago: University of Chicago Press.

Harwood, J. 1997. "National Differences in Academic Cultures: Science in Germany and the United States between the World Wars." In Charle et al. 1997.

Heilbron, J. L., and R. W. Seidel. 1989. *Lawrence and His Laboratory: A History of the Lawrence Berkeley Laboratory.* Berkeley: University of California Press.

Heisenberg, W., and W. Pauli. 1929. "Zur Quantendynamik der Wellenfelder. I." *Zeitschrift f. Physik* 56:1–61.

Heisenberg, W., and W. Pauli. 1930. "Zur Quantendynamik der Wellenfelder. II." *Zeitschrift f. Physik* 59:168–190.

Hentschel, K., and A. Hentschel, eds. 1996. *Physics and National Socialism: An Anthology of Primary Sources.* Basel: Birkhäuser Verlag.

Herbert, G. 1995. *The Complete English Works.* Edited and introduced by A. P. Slater. New York: Knopf.

Herken, G. 1988. *The Winning Weapon: The Atomic Bomb in the Cold War, 1945–1950.* Princeton, N.J.: Princeton University Press.

Hershberg, J. 1988. "Over My Dead Body: James B. Conant and the Hydrogen Bomb." In Mendelsohn et al. 1988.

Hershberg, J. G. 1993. *James B. Conant.* New York: Knopf.

Hertz, D. 1990. "Work, Love and Jewishness in the Life of Fanny Lewald." In Malino and Sorkin 1990, 202–221.

Herz, D. 1988. *Jewish High Society in Old Regime Berlin.* New Haven, Conn.: Yale University Press.

Hewlett, R. G., and O. E. Anderson. 1962. *The New World, 1939–1947: A History of the United States Atomic Energy Commission.* University Park: Pennsylvania State University Press.

Hewlett, R., and F. Duncan. 1972. *A History of the United States Atomic Energy Commission,* vol. 2: *Atomic Shield: 1947–1952.* Washington, D.C.: U.S. Atomic Energy Commission.

Hewlett R. G., and J. M. Holl. 1989. *Atoms for Peace and War: 1953–1961.* Berkeley: University of California Press.

Hill, A. V. 1946. "The Moral Responsibility of Scientists." *Bulletin of the Atomic Scientists* 1(7):3 and 15.

Hirschman, A. O. 1970. *Exit, Voice, and Loyalty: Responses to Decline in Firms, Organizations and States.* Cambridge, U.K.: Cambridge University Press.

Hoddeson, L. (with P. W. Hendricksen, R. Meade, and C. Westfall). 1993. *Critical Assembly: A Technical History of Los Alamos during the Oppenheimer Years, 1943–1945.* Cambridge, U.K.: Cambridge University Press.

Holloway, D. 1980. "Soviet Thermonuclear Development." *International Security* 4/3:193–197.

Holloway, D. 1994. *Stalin and the Bomb: The Soviet Union and Atomic Energy, 1939–1956.* New Haven, Conn.: Yale University Press.

Holton, G. 1984. "Success Sanctifies the Means: Heisenberg, Oppenheimer and the Transition to Modern Physics." In Mendelsohn, ed., 1984, 169–189, and in Holton 1986, 141–162.

Holton, G. 1986. *The Advancement of Science and Its Burden.* Cambridge, U.K.: Cambridge University Press.

Holton, G., and R. S. Morison, eds. 1979. *Limits of Scientific Inquiry.* New York: Norton.

Horkheimer, M., and T. W. Adorno. 1982. *Dialectic of Enlightenment.* Translated by John Cumming. New York: Continuum.

Howe, I. 1976. *The World of Our Fathers.* New York: Harcourt Brace Jovanovich.

Howe, I. 1990. "The Idea of the Modern." In I. Howe, *Selected Writings, 1950–1990,* 140–166. San Diego: Harcourt Brace Jovanovich.

Hoyt, D. C., ed. 1986. *Foucault: A Critical Reader.* Oxford: Basil Blackwell.

Hughes, T. P. 1988. "The Evolution of Large Technological Systems." In W. Bijker, T. P. Hughes, and T. Pinch, eds., *The Social Construction of Technological Systems: New Directions in the Sociology and History of Technology.* Cambridge, Mass.: MIT Press.

Hughes, T. P. 1989. *American Genesis: A Century of Invention and Technological Enthusiasm, 1870–1970.* New York: Viking.

Humboldt, W. von. 1969. *The Limits of State Action.* Edited by J. W. Burrow. Cambridge, U.K.: Cambridge University Press.

Hutchings, Jr., E., ed. 1958. *Frontiers of Science: A Survey.* New York: Basic Books.

International Control of Atomic Energy: Growth of a Policy. 1947. Washington, D.C.: Department of State Publication 2702.

Jaspers, K. 1963. *La bombe atomique et l'avenir de l'homme.* Paris: Buchet-Hassel.

Jensen, K. M., ed. 1991. *Origins of the Cold War.* Washington, D.C.: United States Institute of Peace.

Joint Committee on Atomic Energy. 1951. *Soviet Atomic Espionage.* Washington, D.C.: Government Printing Office.

Jungk, R. 1958. *Brighter than a Thousand Suns.* New York: Harcourt, Brace.

Jungnickel, C., and R. McCormmach. 1986. *Intellectual Mastery of Nature: Theoretical Physics from Ohm to Einstein.* Chicago: University of Chicago Press.

Kant, I. 1947. "What Is Enlightenment?" In Friedrich 1947, 132–140.

Kaplan, M. 1982. "Tradition and Transition: The Acculturation, Assimilation and Integration of Jews in Imperial Germany." *Leo Baeck Institute Yearbook* 27:3–35.

Katz, J. 1986. *Jewish Emancipation and Self-Emancipation.* Philadelphia: Jewish Publication Society.

Keller, E. F. 1992. *Secrets of Life, Secrets of Death.* New York: Routledge.

Kennan, G. F. 1967. "Oppenheimer." *American Scholar,* Fall, 55–56.

Kennan, G. F. 1972. *Memoirs 1950–1963.* Vol. 2. Boston: Little, Brown.

Kevles, D. J. 1995. *The Physicists.* Cambridge, Mass.: Harvard University Press.

Kevles, D. J. 1990. "Cold War and Hot Physics: Science, Security, and the American State, 1945–1956." *Historical Studies in Physical Sciences* 20:2–23.

Keys, D., ed. 1961. *God and the H-Bomb.* Bellmeadows Press with Bernard Geis Associates.

Killian, J. R. 1977. *Sputnik, Scientists, and Eisenhower.* Cambridge, Mass.: MIT Press.

Kistiakowsky, G. B. 1976. *A Scientist at the White House: The Private Diary of President Eisenhower's Special Assistant for Science and Technology.* With an introduction by Charles S. Maier. Cambridge, Mass.: Harvard University Press.

Klein, F. 1979. *Developments of Mathematics in the 19th Century.* Brookline, Mass.: Math Sci Press. (A translation by M. Ackerman of Felix Klein's *Vorlesung über die Entwicklung der Mathematik im 19. Jahrhundert.* Part 1. Berlin: Springer-Verlag, 1928.)

Kline, R. 1996. "Construing 'Technology' as 'Applied Science': Public Rhetoric of Scientists and Engineers in the United States, 1880–1945. *Isis* 87:194–222.

Kloppenberg, J. 1986. *Uncertain Victory: Social Democracy and Progressivism in European and American Thought, 1870–1920.* New York: Oxford University Press.

Kohler, M. J. 1918. *Jewish Rights at the Congress of Vienna (1814–1815) and Aix-La-Chapelle (1818).* New York: American Jewish Committee.

Kraut, B. 1979. *From Reform Judaism to Ethical Culture: The Religious Evolution of Felix Adler.* Cincinnati, Ohio: Hebrew Union College Press.

Krieger, M. 1992. *Doing Physics: How Physicists Take Hold of the World.* Bloomington: Indiana University Press.

Kunetka, J. 1982. *Oppenheimer: The Years of Risk.* Englewood Cliffs, N.J.: Prentice Hall.

Kursunoglu, B. N., and E. P. Wigner, eds. 1987. *Reminiscences about a Great Physicist: Paul Adrien Maurice Dirac.* Cambridge, U.K.: Cambridge University Press.

Kuznick, P. J. 1987. *Beyond the Laboratory: Scientists and Political Activists in 1930s America.* Chicago: University of Chicago Press.

Lal, D. 1993. "A Renowned Cosmic-Ray Physicist: An Obituary of Bernard Peters (1910–1993)." *Current Science* 64 (no. 8):612–614.

Lanouette, W. (with B. Silard). 1992. *Genius in the Shadows.* New York: Charles Scribner's Sons.

Levi, P., and T. Regge. 1989. *Dialogo.* Princeton, N.J.: Princeton University Press.

Lewis, H. W., J. R. Oppenheimer, and S. A. Wouthuysen. 1948. "The Multiple Production of Mesons." *Physical Review* 73:127–140.

Lowen, R. S. 1997. *Creating the Cold War University: The Transformation of Stanford.* Berkeley: University of California Press.

MacKenzie, D. 1996. *Knowing Machines: Essays on Technical Change.* Cambridge, Mass.: MIT Press.

MacKenzie, D. M., and G. Spinardi. 1996. "Tacit Knowledge and the Uninvention of Nuclear Weapons." In MacKenzie 1996, 215–260.

Major, J. 1971. *The Oppenheimer Hearing.* Briarcliff Manor, N.Y.: Stein and Day.

Malino, F., and D. Sorkin. 1990. *From East and West: Jews in a Changing Europe, 1750–1850.* Oxford: Basil Blackwell.

Manuel, F., and F. Manuel. 1979. *Utopian Thought in the Western World.* Cambridge, Mass.: Harvard University Press.

Marshak, R. E., ed. 1966. *Perspectives in Modern Physics. Essays in Honor of Hans A. Bethe on the Occasion of his 60th Birthday, July 1966.* New York: Interscience Publishers.

McClelland, C. E. 1980. *State, Society and University in Germany, 1700–1914.* Cambridge, U.K.: Cambridge University Press.

Mendelsohn, E., ed. 1984. *Tranformation and Tradition in the Sciences. Essays in Honor of I. Bernard Cohen.* Cambridge, U.K.: Cambridge University Press.

Mendelsohn, E., M. R. Smith, and P. Weingart, eds. 1988. *Science, Technology and the Military.* Dordrecht, Holland: Kluwer.

Michaelis, A. R., and H. Harvey. 1973. *Scientists in Search of Their Conscience.* Berlin: Springer-Verlag.

Morgenthau, H. J. 1946. *Scientific Man versus Power Politics.* Chicago: University of Chicago Press.

Morrison, P. 1947. "Physics in 1946." *Journal of Applied Physics* 18:133–152.

Morrison, P. 1988. "Heaven and Earth One Substance: Bernard Peters and the Heavy Primaries." *Current Science* 61, no. 11:740–744.

Mosse, G. L. 1985. *German Jews beyond Judaism.* Bloomington: Indiana University Press.

Murdoch, I. 1992. *Metaphysics as a Guide to Morals.* London: Chatto and Windus; New York: Allen Lane, Penguin Press, 1993.

Muzzey, D. S. 1900. *The Rise of the New Testament.* New York: Macmillan.

Muzzey, D. S. 1902. *Spiritual Heroes: A Study of Some of the World's Prophets.* New York: Doubleday, Page.

Muzzey, D. S. 1907. *The Spiritual Franciscans.* Herbert Baxter Adams Prize essay, awarded December 1905 by the American Historical Association, New York. (Reprint, Washington, D.C.: American Historical Association, 1914.)

Muzzey, D. S. 1911. *An American History.* Boston: Ginn and Co.

Muzzey, D. S. 1915. *Readings in American History.* Boston: Ginn and Co.

Muzzey, D. S. 1918. *Thomas Jefferson.* New York: Scribner's.

Nandy, A. 1988. *Science, Hegemony and Violence: A Requiem for Modernity.* Delhi: Oxford University Press.

Nelson, S., J. R. Barrett, and R. Ruck. 1981. *Steve Nelson, American Radical.* Pittsburgh: University of Pittsburgh Press.

Neumann, J. W. 1916. "Reminiscences of the Workingman's School." *The Standard 2* (May):218–221.

Nietzsche, F. 1957. *The Use and Abuse of History.* New York: Macmillan.

Oleson, A., and J. Voss. 1979. *The Organization of Knowledge in Modern America, 1860–1920.* Baltimore: Johns Hopkins University Press.

Oliver, M. 1992. *New and Selected Poems.* Boston: Beacon Press.

Oppenheimer, J. R. 1926a. "Quantum Theory and Intensity Distribution in Continuous Spectra." *Nature* 118:771–772.

Oppenheimer, J. R. 1926b. "On the Quantum Theory of Vibration-Rotation Bands." *Proceedings Cambridge Philosophical Society* 23:327–335.

Oppenheimer, J. R. 1926c. "On the Quantum Theory of the Problems of Two Bodies." *Proceedings Cambridge Philosophical Society* 23:422–431.

Oppenheimer, J. R. 1927a. "Zur Quantentheorie kontinuierlicher Spektren." *Zeitschrift f. Physik* 41:268–293.

Oppenheimer, J. R. 1927b. "Zur Quantenmechanik der Richtungsentartung." *Zeitschrift f. Physik* 43:27–43.

Oppenheimer, J. R. 1927c. "Bemerkung zur Zerstreuung der α-Teilchen." *Zeitschrift f. Physik* 43:413–415.

Oppenheimer, J. R. 1928. "Three Notes on the Quantum Theory of Aperiodic Effects." *Physical Review* 31:66–81.

Oppenheimer, J. R. 1930a. "Note on the Theory of the Interaction of Field and Matter." *Physical Review* 35:461–477.

Oppenheimer, J. R. 1930b. "On the Theory of Electrons and Protons." *Physical Review* 35:562–563.

Oppenheimer, J. R. 1936. "On the Elementary Interpretation of Showers and Bursts." *Physical Review* 50:389.

Oppenheimer, J. R. 1941. "The Mesotron and the Quantum Theory of Fields." In *Nuclear Physics: University of Pennsylvania Bicentennial Conference.* Philadelphia: University of Pennsylvania Press.

Oppenheimer, J. R. 1945. Address to the Association of Los Alamos Scientists in Los Alamos, November 2, 1945. *Bulletin of Atomic Scientists* 36/6 (June):11–17.

Oppenheimer, J. R. 1946. "Atomic Weapons." *Proceedings of American Philosophical Society* 90/1:7–10.

Oppenheimer, J. R. 1947. *Physics in the Contemporary World.* Cambridge, Mass.: MIT Press.

Oppenheimer, J. R. 1954. *Science and the Common Understanding.* New York: Simon and Schuster.

Oppenheimer, J. R. 1955. *The Open Mind.* New York: Simon and Schuster.

Oppenheimer, J. R. 1958a. *Knowledge and the Structure of Culture.* Vassar College, Widener Library, Harvard University, H5059.58.10.

Oppenheimer, J. R. 1958b. "Talk to Undergraduates." In Hutchings 1958, 239–350.

Oppenheimer, J. R. 1959. "Talk to Undergraduates." *Kansas Teacher,* February 1959, 18–20, 39–40.

Oppenheimer, J. R. 1960. *Some Reflections on Science and Culture*. Chapel Hill: University of North Carolina Press.

Oppenheimer, J. R. 1964. *The Flying Trapeze: Three Crises for Physicists*. New York: Oxford University Press.

Oppenheimer, J. R. 1967. *Oppenheimer*. New York: Scribner.

Oppenheimer, J. R. 1970. *In the Matter of J. Robert Oppenheimer: Transcript of Hearing before Personnel Security Board and Texts of Principal Documents and Letters*. With a foreword by P. M. Stern. Cambridge, Mass.: MIT Press.

Oppenheimer, J. R. 1980. *Letters and Recollections*. Edited by A. K. Smith and C. Weiner. Cambridge, Mass.: Harvard University Press.

Oppenheimer, J. R. 1984. *Uncommon Sense*. Edited by N. Metropolis, G. C. Rota, and D. Sharp. Boston: Birkhäuser.

Oppenheimer, J. R. 1989. *Atom and Void: Essays on Science and Community*. Princeton, N.J.: Princeton University Press. (This collection is a reprint of Oppenheimer 1954 and Oppenheimer 1964.)

Oppenheimer, J. R., and M. S. Plesset. 1933. "On the Production of the Positive Electron." *Physical Review* 44:53–54.

Oppenheimer, J. R., and R. Serber. 1937. "Note on the Nature of Cosmic-Ray Particles." *Physical Review* 51:1113L.

Oppenheimer, J. R., and R. Serber. 1938. "On the Stability of Stellar Neutron Cores." *Physical Review* 54:540L.

Oppenheimer, J. R., and H. Snyder. 1939. "On Continued Gravitational Contraction." *Physical Review* 56:455–459.

Oppenheimer, J. R., and G. M. Volkoff. 1939. "Massive Neutron Cores." *Physical Review* 55:374–381.

Outram, D. 1995. *The Enlightenment*. Cambridge, U.K., and New York: Cambridge University Press.

Pais, A. 1997. *A Tale of Two Continents: A Physicist's Life in a Turbulent World*. Princeton, N.J.: Princeton University Press.

Pauli, W. 1979. *Wissenschaftlicher Briefwechsel*. Edited by K. Hermann, K. von Meyenn, and V. F. Weisskopf, vol. 1: *1919–1929*. New York: Springer-Verlag.

Pauli, W. 1985. *Wissenschaftlicher Briefwechsel*. Edited by K. von Meyenn, vol. 2: *1930–1940*. New York: Springer-Verlag.

Pauly, P. J. 1987. *Controlling Life: Jacques Loeb and the Engineering Ideal in Biology*. Oxford: Oxford University Press.

Peierls, R. F. 1997. "J. Robert Oppenheimer, 1904–1967." In *Atomic Histories*, 46–55. Woodbury, N.Y.: American Institute of Physics.

Phelan, A., ed. 1985. *The Weimar Dilemma: Intellectuals in the Weimar Republic*. Manchester, U.K.: Manchester University Press.

Phillipson, J. 1962. "The Phillipsons, a German Jewish Family, 1775–1933." *Leo Baeck Institute Yearbook* 7:95118.

Pickering, A., ed. 1992. *Science as Practice and Culture*. Chicago: University of Chicago Press.

Pickstone, J. V. 1989. "A Profession of Discovery: Physiology in Nineteenth-Century History." *British Journal of History of Science* 23:207–216.

Price, D. K. 1967. "J. Robert Oppenheimer." *Science* 155 (March 3, 1967). Editorial page.

Price, M. 1995. "Root of Dissent: The Chicago Met Lab and the Origins of the Franck Report." *Isis* 86:222–244.

Proctor, R. N. 1991. *Value-Free Science: Purity and Power in Modern Knowledge.* Cambridge, Mass.: Harvard University Press.

Pyenson, L. 1983. *Neo-Humanism and the Persistence of Pure Mathematics.* Philadelphia: American Philosophical Society.

Pyenson, L. 1985. *The Young Einstein.* Bristol and Boston: Adam Hilger.

Rabi, I. I. 1963. "Science in the Satisfaction of Human Aspiration." In *The Scientific Endeavor: Centennial Celebration of the National Academy of Sciences.* New York: Rockefeller Institute Press.

Rabinow, P., ed. 1984. *The Foucault Reader.* New York: Pantheon.

Rescher, N. 1987. *Forbidden Knowledge and Other Essays on the Philosophy of Cognition.* Dordrecht, Holland, and Boston: D. Reidel.

Rhodes, R. 1987. *The Making of the Atomic Bomb.* New York: Simon and Schuster.

Rhodes, R. 1995. *Dark Sun.* New York: Simon and Schuster.

Richards, I. A. 1952. *Principles of Literary Criticism.* New York: Harcourt, Brace.

Richards, J., ed. 1978. *Recombinant DNA: Science, Ethics and Politics.* New York: Academic Press.

Rieff, P. 1969. *On Intellectuals: Theoretical Studies, Case Studies.* Garden City, N.Y.: Doubleday.

Rigden, J. S. 1987. *Rabi, Scientist and Citizen.* New York: Basic Books.

Ringer, F. K. 1969. *The Decline of the German Mandarins: The German Academic Community, 1890–1933.* Cambridge, Mass.: Harvard University Press.

Ringer, F. K. 1979a. *Education and Society in Modern Europe.* Bloomington: Indiana University Press.

Ringer, F. K. 1979b. "The German Academic Community." In Oleson and Voss 1979, 409–429.

Roberts, J. R., ed. 1979. "*Essential Articles for the Study of George Herbert's Poetry.*" Hamden, Conn.: Archon Books.

Rotblat, J. 1995. "The Post Cold War." Interview with K. Takeuchi. *Asahi Shimbun,* August 5, 1995.

Rouzé, M. 1962. *Robert Oppenheimer: The Man and His Theories.* New York: Paul Eriksson.

Rowe, D. 1986. "Jewish Mathematicians at Göttingen in the Era of Felix Klein." *Isis* 77:422–449.

Sanders, R. 1972. "George Lukács: A Study in European Realism." *Midstream* 18:31–53.

Santillana, G. de. 1968. *Reflections on Men and Ideas.* Cambridge, Mass.: MIT Press.

Sarna, J. D. 1998. "A Great Awakening." In Conser and Twiss 1998.

Schrecker, E. 1986. *No Ivory Tower: McCarthyism and the Universities.* New York: Oxford University Press.

Schrecker, E. 1994. *The Age of McCarthyism: A Brief History with Documents.* Boston: Bedford Books of St. Martin's Press.

Schweber, S. S. 1988a. "The Empiricist Temper Regnant: Theoretical Physics in the United States, 1920–1950." *Historical Studies in Physical and Biological Sciences* 17:17–98.

Schweber, S. S. 1988b. "The Mutual Embrace of Science and the Military: ONR and the Growth of Physics in the United States after World War II." In Mendelsohn et al. 1988, 3–45.

Schweber, S. S. 1989. "Darwin and Herschel: A Study in Parallel Lives. *J. History of Biology* 22/1:1–71.

Schweber, S. S. 1990. "The Young John Slater and the Development of Quantum Chemistry." *Historical Studies in Physical and Biological Sciences* 20, no. 2:339–406.

Schweber, S. S. 1992. "Cornell and M.I.T." In Galison and Hevly 1992, 149–183.

Schweber, S. S. 1995. *QED and the Men Who Made It.* Princeton, N.J.: Princeton University Press.

Seidel, R. 1990. "Books on the Bomb." Essay review. *Isis* 81:519–537.

Serber, R. 1966. Interview by C. Weiner and G. Lubkin. American Institute of Physics, 12.

Serber, R. 1967. "A Memorial to Oppenheimer: The Early Years." *Physics Today* 20, no. 10 (October):35–39. Reprinted in Oppenheimer 1967.

Serber, R. 1992. *The Los Alamos Primer.* Edited and with an introduction by R. Rhodes. Berkeley: University of California Press.

Serber, R. (with R. P. Crease). 1998. *Peace and War: Reminiscences of a Life on the Frontiers of Science.* New York: Columbia University Press.

Shattuck, R. 1996. *Forbidden Knowledge: From Prometheus to Pornography.* New York: St. Martin's Press.

Shea, W. R., and B. Sitter, eds. 1989. *Scientists and Their Responsibilities.* Canton, Mass.: Watson.

Shepley, J. R., and C. Blair, Jr. 1954. *The Hydrogen Bomb: The Men, the Menace, the Mechanism.* New York: David McKay.

Sherwin, M. J. 1975. *A World Destroyed: The Atomic Bomb and the Grand Alliance.* New York: Knopf.

Sherwin, M. J. 1987. *A World Destroyed: Hiroshima and the Origins of the Cold War.* With a new introduction by the author. New York: Vintage Books.

Shils, E. 1956. *The Torment of Secrecy.* Glencoe, Ill.: Free Press.

Shils, E. 1965. "Charisma, Order and Status." *American Sociological Review* 30:199–213.

Shortlaud, M., and R. Yeo. 1996. *Telling Lies in Science: Essays on Scientific Biography.* New York: Cambridge University Press.

Shrader-Frechette, K. 1994. *Ethics of Scientific Research.* Lanham, Md.: Rowman and Littlefield.

Sigurdsson, S. 1991. "Hermann Weyl: Mathematics and Physics, 1900–1927." Ph.D. diss., Harvard University.

Sime, R. 1996. *Lise Meitner: A Life in Physics.* Berkeley: University of California Press.

Sims, J. E. 1990. *Icarus Restrained: An Intellectual History of Nuclear Arms Control, 1945–1960.* Boulder, Colo.: Westview Press.

Smith, A. K. 1965. *A Peril and a Hope: The Scientists' Movement in America, 1945–1947.* Chicago: University of Chicago Press.

Smith, A. K., and C. Weiner, eds. 1980a. *Robert Oppenheimer: Letters and Recollections.* Cambridge, Mass.: Harvard University Press.

Smith, A. K., and C. Weiner. 1980b. "Robert Oppenheimer: The Los Alamos Years." *Bulletin of Atomic Scientists* 36(6):11–17.

Smith, B.L.C. 1992. *The Advisers: Scientists in the Policy Process.* Washington, D.C.: Brookings Institution.

Snow, C. P. 1959. *The Two Cultures and the Scientific Revolution.* Cambridge, U.K.: Cambridge University Press.

Söderqvist, T. 1996. "Existential Projects and Existential Choice in Science: Science Biography as Edifying Genre." In Shortland and Yeo 1996, 45–84.

Solomon, J. J. 1973. "Science and Scientists' Responsibility in Today's Society." In Michaelis and Harvey 1973.

Sommerfeld, A. 1929. *Atombau und Spektrallinien: Wellenmechanischer Erganzungsband.* Braunschweig: F. Vieweg.

Sommerfeld, A., and H. A. Bethe. 1933. "Elektronentheorie der Metalle." *Handbuch der Physik*, vol. 24/2. Berlin: Geiger/Scheel.

Sorkin, D. 1983a. "Wilhelm von Humboldt: The Theory and Practice of Self-Formation (*Bildung*), 1791–1810." *J. History of Ideas* 44:55–73.

Sorkin, D. 1983b. "Ideology and Identity: Political Emancipation and the Emergence of a Jewish Subculture in Germany, 1800–1848." Ph.D. diss., University of California, Berkeley.

Stern, P. M. (with the collaboration of H. P. Green). 1969. *The Oppenheimer Case.* New York: Harper and Row.

Strauss, L. L. 1962. *Men and Decisions.* Garden City, N.Y.: Doubleday.

Strier, R. 1983. *Love Known.* Chicago: University of Chicago Press.

Sweet, P. R. 1978. *Wilhelm von Humboldt: A Biography,* vol. 1: *1767–1808.* Columbus: Ohio State University Press.

Sweet, P. R. 1980. *Wilhelm von Humboldt. A Biography,* vol. 2: *1808–1835.* Columbus: Ohio State University Press.

Sylves, R. 1987. *The Nuclear Oracles: A Political History of the General Advisory Committee of the Atomic Energy Commission, 1947–1957.* Ames: Iowa State University Press.

Szasz, F. M. 1984. *The Day the Sun Rose Twice.* Albuquerque: University of New Mexico Press.

Szilard, L. 1978. *Leo Szilard: His Version of the Facts.* Edited by S. R. Weart and G. Weiss Szilard. Cambridge, Mass.: MIT Press.

Tamm, I. 1930. "Über die Wechselwirkung der freien Elektronen mit Strahlung nach der Diracschen Theorie des Elektrons und nach der Quantenelektrodynamik." *Zeitschrift f. Physik* 62:545–568.

Teller, E. 1955. "The Work of Many People." *Science* 121 (February 25):267–275.

Teller, E. (with A. Brown). 1962. *The Legacy of Hiroshima.* Garden City, N.Y.: Doubleday.

Teller, E. 1983. "Seven Hours of Reminiscences." *Los Alamos Science* 4/1: 190–196.

Tillich, P. 1961. "The Power of Self-Destruction." In Keys 1961, 35–36.

Tolman, R. C. 1917. *The Theory of the Relativity of Motion*. Berkeley: University of California Press.

Tolman, R. C. 1934. *Relativity, Thermodynamics and Cosmology*. Oxford: Clarendon Press.

Traweek, S. 1992. "Narrative Strategies in Science Studies and among Physicists in Tsukuba." In Pickering 1992, 429–462.

Truslow, E. C., and R. C. Smith. 1983. *Project Y: The Los Alamos Story*, Part 2: *Beyond Trinity*. Los Angeles: Tomash.

Uehling, E. A. 1935. "Polarization Effects in the Positron Theory." *Physical Review* 48:55–63.

Ulam, S. 1976. *Adventures of a Mathematician*. New York: Scribner.

Ulam, S. 1987. "Vita." *Los Alamos Science*, Special Issue 1987, 8–22. (Excerpts from Ulam 1976.)

Vendler, H. 1975. *The Poetry of George Herbert*. Cambridge, Mass.: Harvard University Press.

Visvanathan, S. 1988. "Atomic Physics: The Career of an Imagination." In Nandy 1988, 113–166.

Waller, I. 1930. "Bemerkung über die Rolle der Eigenenergie des Elektrons in der Quantentheorie der Strahlung." *Zeitschrift f. Physik* 62:673–676.

Walter, M. L. 1990. *Science and Cultural Crisis: An Intellectual Biography of Percy Williams Bridgman (1882–1961)*. Stanford, Cal.: Stanford University Press.

Wang, J. 1992. "Science, Security, and the Cold War: The Case of E. U. Condon." *Isis* 83:238–269.

Wang, J. 1999. *American Science in an Age of Anxiety: Scientists, Anticommunism, and the Cold War*. Chapel Hill, N.C.: University of North Carolina Press.

Weart, S. 1988. *Nuclear Fears: A History of Images*. Cambridge, Mass.: Harvard University Press.

Weber, M. 1918. "Science as Vocation." In Weber 1946, 125–156.

Weber, M. 1946. *From Max Weber: Essays in Sociology*. Translated and edited by H. H. Gerth and C. Wright Mills. Oxford: Oxford University Press.

Weber, M. 1947. *The Theory of Social and Economic Organization*. New York: Oxford University Press.

Weber, M. 1968. *On Charisma and Institution Building*. Selected Papers. Edited with an introduction by S. N. Eisenstadt. Chicago: University of Chicago Press.

Weinberg, S. 1995–1996. *The Quantum Theory of Fields*. 2 vols. Cambridge, U.K., and New York: Cambridge University Press.

Weiner, C. 1969. "A New Site for the Seminar: The Refugees and American Physics in the Thirties." In Fleming and Baylin 1969, 152–189.

Weiner, C., ed. 1977. *Proceedings of the International School of Physics "Enrico Fermi." Varenna, Italy. Course 57: History of Twentieth-Century Physics*. New York: Academic Press.

Weisskopf, V. F. 1934a. "Über die Selbstenergie des Elektrons." *Zeitschrift f. Physik* 89:27–39.

Weisskopf, V. F. 1934b. "Berichtung." *Zeitschrift f. Physik* 90:817–818.

Weisskopf, V. F. 1954. "Science for Its Own Sake." *Scientific Monthly* 72:133–135.

Weisskopf, V. F. 1970. "Intellectuals in Government." In Allen 1970, 25–29.

Wigner, E. P., ed. 1946. *Physical Science and Human Values.* Princeton University Bicentennial Conference on the Future of Nuclear Science. Princeton, N.J.: Princeton University Press.

Wigner, E. P. 1995. *Collected Works.* Edited by A. S. Wightman. Princeton, N.J.: Princeton University Press.

York, H. 1989. *The Advisors: Oppenheimer, Teller and the Superbomb.* Stanford, Cal.: Stanford University Press.

Young, W. H. 1928. "Christian Felix Klein, 1849–1925." *Proceedings Royal Society London* A 121:1–19.

Ziegler, C. 1988. "Waiting for Joe-1: Decisions Leading to the Detection of Russia's First Atomic Bomb Test." *Social Studies of Science* 18:197–229.

INDEX